THE FUTURE OF U.S. ROCKETRY

by

Edward Hujsak

Mina-Helwig Company
P.O. Box 1292
La Jolla, CA 92038-1292

Illustrations: Edward Hujsak, author

Library of Congress Cataloging-in-Publication Data

Hujsak, Edward, 1925-
The Future of U.S. Rocketry / by Edward Hujsak.
p. cm.
Includes bibliographical references and index.
ISBN 1-886133-01-8 (alk paper)
1. Rocketry--United States--History. 2. Rocketry--United States--Technological Innovations. 3. Launch Vehicles (Astronautics)
I. Title. II Title: The Future of U.S. Rocketry

TL781.8.U5H85 1994b
621.43'56'0973--dc20 94-35384
CIP

International Standard Book Number (ISBN) 1-886133-01-8

Published and Distributed by: Mina-Helwig Company
P.O. Box 1292
La Jolla, CA 92038 -1292
Tel: 619.457.9738
FAX: 619.457.9738

Printed in the United States of America

1 2 3 4 5 6 7 8 9 0

We have been here since the beginning of time. Just recently, through some magnificent trick of entropy, we have taken the form of sentient beings - for the blink of an eye, as time goes. Born into a chorus of living things, we are thoughtful, caring, communicative, intuitive, inventive, adventurous, capable of stupendous deeds. How sad, if in the course of events, humanity opts to hunker down, passing up a signal opportunity to expand beyond Earth's boundaries. Entropy rules. One by one we return to matter, where we will remain 'til the end of time.

To Joy, who held the fort while
I was away playing with rockets

CONTENTS

1

A Half Century of Rocketry

1-1 The Space Race

Modern rocketry spans a mere fifty years. The pioneer work accomplished by Robert Goddard in the United States was brought into sharp focus with the development of the V-2 rocket by the Germans during World War II. Remnants of their capability, including engineers and hardware, were commandeered by both the United States and Soviet Russia when Germany was defeated. At the end of the war, each nation had the benefit of V-2 technology and a core of German engineers to kick off their own programs in rocketry.

Thus, the end of the war saw the beginning of a race between the two powers to develop formidable fleets of opposing missiles armed with nuclear warheads.

Early U.S. programs were directed toward development of both intercontinental ballistic missiles (ICBM's) and intermediate range ballistic missiles (IRBM's). The first of the ICBM's, developed by the Air Force, were General Dynamics' Atlas and Martin Marietta's Titan I. Titan I did not reach operational service. Several versions of Atlas were developed, which were emplaced in above-ground launchers and in silos at a number of sites within the continental U.S. The first IRBM's were the Redstone, Jupiter, and Thor. Redstones and Jupiters were developed at Redstone Arsenal by engineers led by Werner von Braun and others who built the German V-2. Production models were built by Chrysler Corporation. Redstones were deployed at various locations by the Army. Jupiter was built in competition with the Air Force Thor. Eventually, the Air Force was assigned responsibility for deployment of both vehicles. Field installations were in Turkey, Italy, Federated German Republic, and Great Britain. These systems were withdrawn after the Cuban missile crisis in 1963.

Scientists and engineers were already aware that the new ballistic missiles possessed the capability to carry spacecraft into Earth orbit and beyond into the solar system. In 1955, development work was begun on the Vanguard rocket to conduct a series of orbiting scientific investigations. This vehicle was beset by problems and was overtaken by a Jupiter C, which performed the first successful orbital launch. Vanguard, designed to place a mere 20 pounds into orbit, was dropped from use after two successful flights. Space launches advanced rapidly after the first flight by Jupiter C. Modern space rocketry is a product of the technology that was developed to build ballistic missiles.

Initially, because early models of nuclear warheads were heavy, the ICBM's were correspondingly large vehicles. The first Atlas was a five-engine device, 12 feet in diameter. In the course of events, advanced versions of nuclear warheads were developed which were much lighter. This led to a redesign of U.S. missiles, reducing their size and lift capability. Atlas was downsized to three engines and ten feet in diameter. But the Soviets were slower in achieving lightweight warheads. This quirk of development led to their construction of larger launch vehicles than those built in the U.S. As a result, in the early days of modern rocketry, the Soviets were able to launch larger spacecraft into orbit than the U.S., a factor that led directly to their achievement of the first manned space flight.

Early Atlas ICBM before downsizing for lighter warheads was twelve feet in diameter. (Gen Dyn)

A giant leap forward was initiated by President John F. Kennedy's announcement in May of 1961 that the nation would undertake the challenge of placing humans on the moon in that decade. The project was viewed as critical to the goal of assuring U.S. dominance in the new field of space exploration. From the time the Soviets launched Sputnik in 1957, the moon had been viewed by many Americans as the next logical goal in what was rapidly becoming a two-nation space race. NASA had begun serious studies of a lunar landing in 1959, two years before President Kennedy's announcement. By 1960, industrial concerns were on contract to examine key aspects of the enterprise. America already had a program called Project Mercury to place a man into orbit and recover him safely.

As some expected would occur, the Soviets took up the challenge under a program cloaked in secrecy. But their technologists did not measure up to the task. They were unable to achieve timely production of reliable rockets necessary for the mission. In contrast, in full public view, U.S. industry, under a strong NASA leadership, performed magnificently in producing a system that exceeded all expectations for the moon venture called Apollo.

Apollo's success was due not only to the development momentum that the U.S. had acquired through its ICBM programs, but also to a stellar group of engineers and administrators who led the program. The ICBM programs were a vast training ground. They developed the disciplines that were necessary to make the Apollo program work.

Apollo settled the space race. From then on there was no question that U.S. technological prowess in space exploration had no match. The question was: where do we go from here? There was no shortage of options. Preliminary design engineers at all the major aerospace firms, scientists at many universities, as well as NASA and DoD engineers and scientists were busy throughout the 1960's developing program concepts that ranged from aerospace planes to manned planetary missions. Apollo was just the first step in a rapidly expanding area of human activity.

1-2 The Aftermath of Apollo

The last lunar mission was completed in December, 1972. In six successful flights a dozen men took the first tentative steps toward exploration of the moon. It was a remarkable achievement, and one that would be hard to follow.

Skylab and several ambitious unmanned planetary explorers were already underway as new projects. NASA moved quickly to capitalize on the Saturn launch systems that launched Apollo, creating the first manned space station, called Skylab. The Space Station was built about a shell structure common to the propellant tanks used on the Saturn SIVB upper stage, which was the third stage of Apollo's Saturn V launch vehicle. Skylab supported Earth and space research missions and accommodated crews for up to 84 days in orbit. It was a highly successful but short-lived program. The station was abandoned in 1974.

With the conclusion of the Apollo program, enthusiasm for further manned exploration of space waned within the administration and in the Congress, resulting in a curtailment of funding for manned programs. A bright spot was growth of rocketry in the commercial sector. Communications satellites were beginning to demonstrate high income potential. Growth in this area proceeded largely through efforts of private concerns and international consortia. They required only that the government provide them with the rockets to place their spacecraft into geosynchronous orbit.

Rapid progress in space exploration is inhibited by the high cost of launch vehicles. Rockets are special insofar as mechanical devices are concerned because they are required to operate at the upper limit of the capability of the materials from which they are constructed. Whereas normal industrial equipment is built with a factor of safety of four, rockets are designed to operate with a factor of safety approaching one. This is strictly a consequence of the necessity to make them as light as possible, maximizing their ability to carry payload. Another contributor to high cost is low production rate. A visitor to a launch-vehicle factory production line is unlikely to encounter assembly workers. The total time required to assemble a launch vehicle is only a few hundred hours. But the time elapsed on the assembly line might be six months. Thus, each vehicle carries a large share of the overhead costs incurred in keeping the factory in operation.

While contractors are constantly spurred by procuring agencies to lower launch-vehicle costs, it is doubtful that in a given class of vehicles the costs can ever be lowered by as much as fifty percent. The rationale for this rests in the observation that the nonmarket economies of China and Russia can offer space launches at about half the cost of U.S. or

European launch vehicles. Wages in those countries are trivial as compared to U.S. wages. Although some improvements in hardware producibility can be achieved to lower cost, the fundamental problem still resides in high labor rates and low production rates.

For a short time it was believed that commercializing the launch vehicle industry would result in a significant lowering of costs. High cost was then attributed to the burdensome specifications and controls imposed by the government on the manufacturers. Every component on a launch vehicle is accompanied by an extensive quality control paper trail that originates when the part first enters the manufacturing cycle. Parts traceability is vital for support of flight failure analyses. When manufacturers began to produce commercial launch vehicles, they discovered that few of the government imposed controls could be discarded, so anticipated large cost reductions did not occur. Flight failures are traumatic occurrences in the launch industry. But they will remain common occurrences until more robust systems are developed.

The new idea was that rockets should be reusable, operating like aircraft to perform hundreds of missions over their lifetimes. Once reusability was achieved, the cost of launching payloads into orbit would drop dramatically, or so it was believed. Years before the Apollo project ended, NASA had begun working on the problem, deeming it critical for future space operations. There would be a "Space Shuttle" which would operate like a truck, hauling payloads and transfer stages of various weights and configurations to and from Earth orbit at a fraction of the cost of expendable launch vehicles. The Space Shuttle would have a life cycle of hundreds of flights, and expendable launch vehicles would be obsoleted. Only a half dozen or so Shuttles would be needed to perform the whole gamut of planned missions.

U.S. industries immediately jumped aboard, recognizing in the process that they would be undercutting the business base in expendable launch vehicles they had worked so hard to develop. The program was artfully sold to the Congress, to the degree that the DoD was pressed to abandon its fleet of expendable launch vehicles and engage in a cooperative effort with NASA to develop the new system. NASA, in turn, as the primary supplier of launch vehicles for commercial spacecraft, began a curtailment of the supply of launch vehicles for that purpose. As the sole provider of launch vehicles to the commercial trade, it was a simple matter to shunt commercial payloads from expendable launch vehicles to Space Shuttle manifests.

Several years were spent performing design studies for the Shuttle system. Shuttle concepts that featured flyback boosters as well as a returnable orbiter were favored from technical and operational standpoints. Congress, however, balked at the price tag and forced NASA to come up with a compromise configuration. The cost of development of two fly-back vehicles, the booster and the orbiter, was too high. The configuration that was finally selected included a recoverable orbiter fitted with a crew compartment and cargo bay, a jettisonable propellant tank called the ET, which supplied oxygen and hydrogen to the orbiter's main engines, and a pair of solid-propellant booster rockets that were to be recovered and processed for reuse after landing in the ocean. Curiously, the commitment to build a Space Shuttle that would be a carrier for all payloads remained untouched, with the result that the expendable launch vehicle industry began a shutdown of its production lines.

As the Space Shuttle entered operation in the early 1980's, it became evident that the promise of a do-all launch system was ill-founded. Development and operations problems that beset the program from the beginning never abated, raising havoc with Shuttle launch schedules. Users began to worry that NASA would never be able to launch their payloads on schedule. Moreover, even though NASA offered users bargain prices to launch their payloads, the high overall costs associated with the program gave early indication that the system would never be competitive with expendable launch vehicles, much less undercut them by substantial margins as originally predicted.

A far more serious concern in the Department of Defense was that a problem with the Space Shuttle could cause it to be grounded for extended periods, leaving the nation without a launch system. This concern proved to be justified by the Challenger accident in January of 1986, which delayed further Shuttle flights for two years. The prospect of extended stand-downs led Air Force Secretary E.C. Aldridge to present a case to Congress to restart the expendable launch vehicle production lines. His proposal was approved, and by 1984 a new era had begun for the thirty year old Atlas, Delta, and Titan rocket systems.

The government was no longer interested in being the procuring agency for launching commercial payloads, freeing the contractors to set about pursuing their own markets and supplying them as they saw fit. The government was cooperative, however, in allowing the use of existing launch facilities for commercial purposes. General Dynamics, McDonnell Douglas and Martin Marietta embarked on manufacturing facilities and hardware improvement programs to prepare for the worldwide market in commercial launches. Several companies operating as purely commercial enterprises sprang up to serve the small payload market.

The Department of Defense and NASA continued to procure launch vehicles for their own purposes. Titan IV was configured solely to serve Air Force missions. Atlas and Delta were classified as medium launch vehicles (MLV). From time to time the DoD has contracted for production of quantities of these vehicles. MLV I and MLV III are Delta rockets, while MLVII consists of Atlas with a Centaur upper stage.

1-3 Commercial Rocketry - A Viable Industry?

During 1984, in a move to assist the budding commercial launch industry, the Reagan administration created the Office of Commercial Space Transportation within the Transportation Department. Its duties included review of commercial launch missions to ensure public safety as well as other functions that were intended to help nurture and protect a new but risky enterprise. The commercial launch industry seemed to fit hand in hand with U.S. government policy that forbade export of U.S. built spacecraft to foreign launchers. Unfortunately, only a few years passed before the government began to undermine its own policy, first by granting the Chinese government rights to launch as many as nine U.S. built commercial satellites on its Long March II rocket, and more recently granting the same privilege to Russia to launch a U.S. built satellite on the Proton rocket.

Foreign launchers pose a significant threat to the U.S. commercial launch industry. The problem is whether commercial launch operations in the U.S. can compete against nonmarket economies. Given free rein, Chinese and Russian rockets will soon dominate the business, offering services at roughly half the cost of U.S. launchings. The business base is thin. In the years between 1996 and 2002 there will be only fifteen to eighteen commercial launches annually. Forty percent will be launched by Europe's Ariane. The remainder, if shared among Atlas, Delta, Long March II and Proton, and possibly the Japanese H-2, will seriously threaten sustainability of U.S. commercial launch efforts.

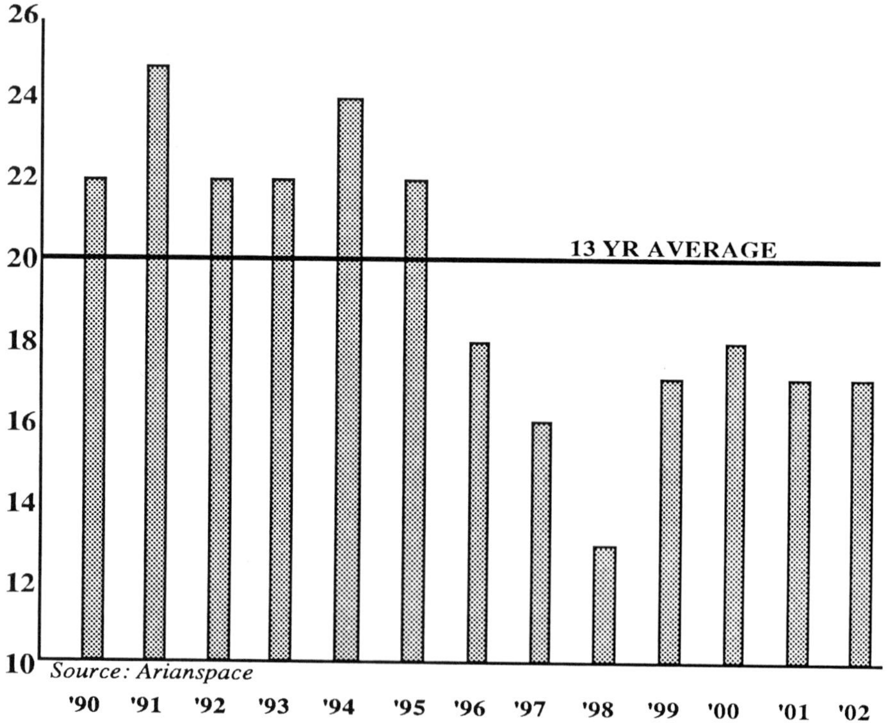

The launch vehicle industry faces a diminishing commercial market in the last half of this decade.

Other factors darken the scene for the commercial launch industry in the coming years. Spacecraft design has progressed to the point that around 60,000 telephone lines can be carried and a life expectancy of 10 to 12 years is assured. But models now under development will have a capacity of up to 100,000 telephone lines and will operate for 16 years. This trend in satellite design corresponds directly to fewer launches.

Advances in digital compression also work to increase the efficiency of satellites. Compression technology can squeeze several video channels into a single satellite transponder that previously carried only one channel.

Yet another influence is the emerging fiber optic network that will be emplaced in the developed nations in the near future. The enormously higher bandwidth inherent in optical transmission, and improved accuracy of transmission that is so important to computer networking, forecasts a takeover of large segments of the communications market by ground-based optical systems. In this event, the remaining major market for communications satellites will be the underdeveloped countries, as their only affordable option.

At present, U.S. commercial space activities are experiencing rapid growth, fueled by requirements for new satellite services. Information compiled by the Commerce Department's Office of Space Commerce shows that total participation in remote sensing, commercial launches, satellite ground equipment, satellite services and commercial satellites has tripled since 1988. However, the U.S. share of commercial launches is declining from 51 per cent of the market in the 1980's to a firm 21 per cent in the 1990's, with 38 percent as yet undetermined. If U.S. manufacturers succeed in winning half of the unassigned 38 per cent, their share for the 1990's will drop to 40 per cent.

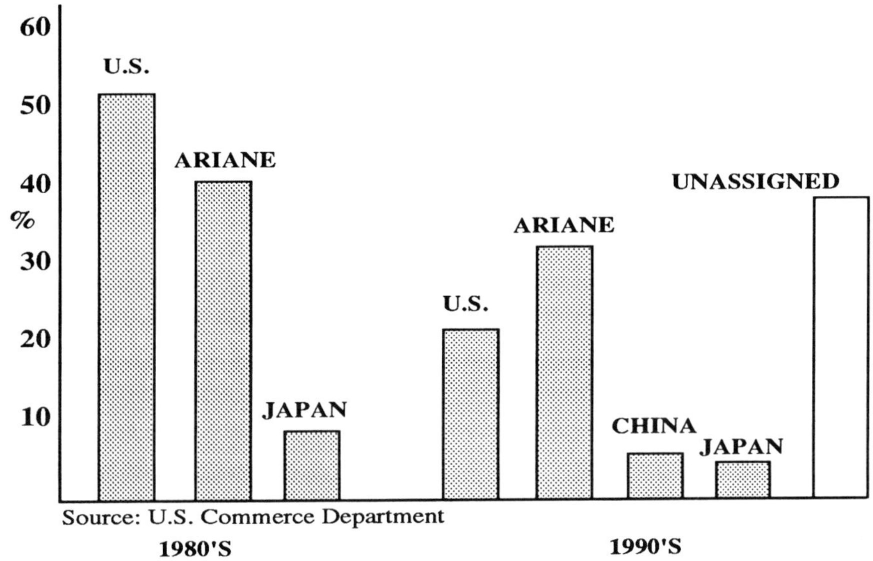

Foreign competition has reduced the U.S. share of the commercial launch market

Still further exacerbation of the commercial launch industry will occur if partnerships such as the recent joint venture between Lockheed and Khrunichev, producers of the Russian Proton rocket, are permitted to compete for U.S. built satellite launches. By investing in Motorola's mobile communications satellite venture, named Iridium, Khrunichev has already achieved an inroad by providing three launches. This system constitutes as many as sixty-six small satellites operating in low-earth orbit, providing continuous worldwide coverage for mobile telecommunications. The Proton rocket is capable of launching the Iridium satellite in groups of seven at a time.

As builder of the Iridium satellites for Motorola, Lockheed has maneuvered into a position of considerable strength with respect to influencing selection of foreign launch vehicles for space launches. The Lockheed Khrunichev accord is viewed by Lockheed officials as having the potential to compete regularly for Western commercial contracts. In this sense, it will be a formidable threat to Europe's Ariane launcher as well as U.S. rockets.

To sum up, the fledgling U.S. commercial launch industry could easily become the victim of conflicting government policies, or suffer extinction by default due to absence of protective measures to ensure preservation of this important industry. To its credit, the Government is insisting in its agreements that China and Russia conform to pricing standards in their bidding for U.S. satellite launches. But how this can work under current methods of proposal submittal, in which price is the determining factor for contract award, seems to defy understanding.

One area of growth in the commercial arena is in launch vehicles for lightweight satellites. Advances in hardware miniaturization has seen the emergence of fully functional spacecraft that are a fraction of the weight of conventional communications satellites. They can therefore be launched in groups of three to six on medium payload rockets like Atlas and Delta, or they can be launched singly on smaller rockets. Small launch vehicles will play a role in the Iridium and similar systems when the time arrives to launch replacement satellites.

Some encouragement for commercial launchers comes from a set of guidelines issued by the White House in February of 1991. The guidelines encourage agencies to serve as anchor tenants for space business ventures. The document also allows unused government assets or devices to be made available to the private sector as long as the economic impact of the transfer is taken into account. The provision is aimed at surplus military rockets that can be used commercially. A large supply of unused rocket stages already exists in government storage as a result of obsoletion of systems and the end of the Cold War. Cheap launches could be had by using this resource, but implementation will have a direct, adverse affect on new rocket production.

Rocketry exists at the behest of missions. Missions, in turn, are derived from requirements. A requirement can originate in many places. It can be created by a single engineer engaged in operational analysis or by groups of scientists at JPL or analysts at the Rand or Aerospace Corporations. Once a requirement is defined and validated, the mission can be designed. Most will undergo preliminary design studies to investigate and select ways of doing the job. This will be followed by further analysis and planning exercises within the originating agency before seeking approval at higher levels. If approval seems likely, a second round of studies, usually performed by industry, will be undertaken. Eventual approval is obtained in the Congress or other sponsoring party, and the project is funded and underway. NASA and the Defense Department each have fixed routines under which a new project progresses through the approval process.

A requirement can be validated by Presidential imperative. It is the fast track way to get a mission underway. When President Kennedy announced that the United States was going to send humans to the moon, he validated a requirement. NASA already had done the preliminary studies that assured him that the task could be accomplished. President Reagan validated a requirement when he announced the Strategic Defense Initiative (SDI)). The SDI proposed to build a protective space umbrella of sensors and attack systems that would do sentry duty against ballistic missile attacks on the U.S. In this instance, Edward Teller and other champions of the idea at Lawrence Livermore Laboratories were able to win the President's attention by direct approach. Both Apollo and SDI subsequently experienced little trouble in obtaining congressional approval.

When the requirement is established, the mission definition can be seriously pursued. Launch vehicle requirements constitute a part of the mission definition. Usually designers seek to constrain spacecraft weight and size to the lift capability of existing launch vehicles. If this is not possible, modifications to the launch vehicles, or even completely new designs will be specified.

Only rarely in the short history of rocketry has an all-new space launch vehicle or family of launch vehicles been specified. Apollo is one case. There was no existing capability to lift the heavy payloads required for the moon mission, and so the Saturn rocket family was created for Apollo. The Space Shuttle is another case. Its objective was to provide a low-cost launch capability to replace expendable launch systems. The Strategic Defense Initiative covered so many options that it was impossible to arrive at launch vehicle requirements. Most other mission requirements in the past fell within the capabilities of Atlas, Delta, and Titan rockets.

Industry has been reluctant to undertake launch systems modernization or creation of new launch systems through its own initiatives, relying instead on government funding and immediate commercial requirements to set the pace of development. Small launch vehicles are the sole exception, as entrepreneurs have stepped up to the challenge of participating in the space launch business through private initiative. Small vehicle development is no longer driven by specific requirements. The industry can predict with some confidence a rapidly developing market for a variety of small satellites. Larger launch vehicles are viewed on the other hand by the aerospace industry as too expensive and risky to develop through private investment.

An indication that this may change is indicated by the recent formation of an alliance among five major aerospace companies to evaluate the market potential for a future low-cost transportation system. The five companies - Boeing, General Dynamics, Lockheed, Martin Marietta and Rockwell International - will explore the prospective market for a launch system that would support traditional satellite deployment as well as possible new markets such as power generation in space and use of space for special movie film effects in zero gravity. Launch vehicle studies would begin only after the market analysis has been completed. But the companies cannot be complimented for having taken up the initiative on their own. The market study is being funded by a $3M NASA grant, the terms of which place NASA in the role of "Industry Facilitator."

Over the past eight years, extensive DoD and NASA efforts have been directed toward obtaining development authorization for a modern launch system to replace Atlas, Delta, and Titan. Better reliability, system resiliency, improved responsiveness, lower cost, and greater payload carrying capability have all been cited as justifications for the new venture. Here is a case where the requirements are nonspecific, and where the mission is all-encompassing, a situation that the Congress has difficulty comprehending. It is difficult for proponents to satisfactorily answer the first question, " Doesn't the existing system work?"

The search for a new launch vehicle system began in 1987 with a "Space Transportation Architecture Study (STAS)," performed by General Dynamics and Boeing under joint funding by DoD and NASA. The Architecture Study found that the largest single payload planned for delivery into low-earth orbit was a spacecraft with its high energy upper stage, for a total weight of 70,000 pounds. The study suggested that a new launch system should have a capacity of 100,000 pounds or more to low-earth orbit.

Over the ensuing years, the DoD and NASA funded industry studies to define what was first termed the Advanced Launch System (ALS) and later the National Launch System (NLS). Both programs were directed toward developing a heavy-lift capability. In the beginning, the requirement for heavy-lift vehicles was drawn from SDI plans which were far from firm. When these plans were dropped, the lift capability for the NLS was reduced. Concurrent with the NLS studies, NASA was pursuing its own ideas for a heavy-lift vehicle with the Shuttle "C" concept, which was based on Space Shuttle components. Still, no requirement for heavy-lift capability had appeared beyond what was achievable with Titan IV. In 1992 Congress finally terminated work on the NLS, citing high cost of development and absence of a firm mission.

Nevertheless, DoD has not given up its quest for a new launch vehicle. A 1992 study for the White House Space Council, instigated by Vice President Quayle, called for developing a new family of launch vehicles with a launch capability that spans the existing band from Delta to Titan IV. The "Spacelifter " rockets described in the report would reduce the cost of putting payloads into orbit, increase reliability, and improve ability to compete in the commercial market. It remains to be seen whether Congress will eventually approve development of a new system for these reasons. A possible administration argument for the project could be to revitalize the aerospace industry segment, which has been severely affected by the end of the Cold War. The project has its opponents in the Space Shuttle program, who view it as a threat to that system.

Indeed, Spacelifter has few supporters among rocket manufacturers. The rocket industry has been carefully entrenched through geopolitical measures prompted by self-preservation instincts. Segments of the industry are distributed across key states in a manner that engenders strong support by key members of Congress. Appearance of Spacelifter in the congressional approval cycle would no doubt find subtle opposition that could be traced to manufacturers' reluctance to abandon a sure thing for a chance to bid on something new.

In retrospect, it could be said that the DoD committed a strategic error when approval to restart Atlas, Delta, and Titan production was secured from Congress in the early 1980's. The message to Congress at that time might well have been: "This is an expedient action to avert loss of launch capability in the event of a problem with the space shuttle. But these vehicles are old designs and will be expensive to procure and operate. Authorization is also needed to concurrently develop a modern launch system to replace them as early as possible." *That was the requirement.*

With time, the requirement became fuzzy. It is a situation analogous to wanting to buy a new car despite the fact that the old one works well. Today, with the industry having made large investments to improve producibility and performance of their own products, Congress can justifiably question whether the requirement exists at all.

1-5 A Space Program in Disarray

It is no surprise that an aura of distrust has developed regarding space programs, when it appears that they are suffering from an absence of overall direction and purpose. The feeling is a consequence of the numerous program starts and terminations that have occurred in the past few years.

In July, 1989, the 20th anniversary of the first Apollo landing on the moon, President Bush announced the "Space Exploration Initiative" (SEI), a fast paced program that included a return to the moon to stay, and to land humans on Mars. The new space policy called for expanding human presence and activity beyond Earth orbit into the solar system. It included objectives of obtaining scientific and economic benefits of space for the American people, encouraging private sector participation in space, improving the quality of life on Earth, strengthening national security, and promoting international cooperation in space.

In August of 1989 NASA began an extensive review to summarize the strategies and technologies needed to support the program. Vice President Dan Quayle, Chairman of the National Space Council, requested NASA to "cast a net widely to find the most innovative ideas in the country." This formed the basis of the "Outreach Program." Much of the solicitation of ideas was done through a contract issued to Rand Corporation and through a survey conducted by the American Institute of Aeronautics and Astronautics.

Vice President Quayle then formed what came to be known as the "Synthesis Group," under the direction of retired Air Force General Thomas Stafford. The group was chartered to provide two or more architectures, technology requirements, and the early accomplishments needed to support the SEI.

The study produced four architectures and listed ten recommendations. Four recommendations dealt with administration and strategy, one recommended inclusion of SEI objectives in the NLS program, and three specified technology development in nuclear power, nuclear propulsion, and life sciences. The last two called for education as a principal SEI theme and a continuation of the Outreach Program. The Synthesis Group report was released in May of 1991 at approximately the same time that the President announced a shift in SEI policy.

Congressional reluctance to act on the Space Exploration Initiative prompted President Bush to shift his strategy toward more modest objectives that focused on technology and were not constrained by schedule. This seemed to be in accord with implied "go slow" recommendations made by the Synthesis Group. Over a period of two years, only seven million dollars was authorized for the program, most of it earmarked for studies.

In July of 1990 Vice President Quayle and NASA Administrator Richard Truly formed the "Advisory Committee of the Future of the U.S. Space Program" to examine civil space goals, programs and management. The committee was chaired by Norm Augustine, chairman and CEO of Martin Marietta Company. Hence this came to be known as the Augustine Committee. The Committee released its findings in December of 1990.

The panel's most extensive findings dealt with the Space Station. It recommended a radical redesign of the station to focus on life sciences instead of materials processing. The panel also recommended development of a new launch system, rejecting upgrades to existing systems, including the Space Shuttle. It called for science and new launch-vehicle development to take priority over the Space Exploration Initiative. It also recommended significant organizational changes in NASA's program management. Three years later it appears that the first and last of these recommendations appear to have been heeded, but for reasons other than recommended by the Augustine Committee. The Committee's recommendations regarding a new launch vehicle fell on deaf ears.

A third investigation to examine the future of U.S. space launch capability was commissioned by Vice President Quayle in July of 1992. The Task Group, headed by former Secretary of the Navy E.C. Aldridge, also came out with strong recommendations for development of a new launch vehicle, now dubbed "Spacelifter." Unfortunately, the report was delivered in November of 1992, in a period between administrations. This important body of work fell homeless, with no one to champion its cause.

Thus, three independent thrusts at preparing composite pictures of the nation's future space programs were completed over a period of two years. Each committee called upon the best available talents to provide technical and programmatic inputs. All called upon the government to begin development of a modern launch system. Although compliments flowed regarding the depth and thoughtfulness of the studies, none appear to have been taken seriously.

U.S. space programs still continue with a momentum of their own. The commercial market is small. Its future is threatened by low cost foreign rockets. The DoD and NASA have their own agendas which provide a steady demand for launch vehicles from the existing stable. Small satellites, and resulting demand for small launch vehicles, shows growth in this area in the future. Notwithstanding, the future of space exploration for the nation lies largely undefined. The Congress is unwilling to respond to the recommendations of the best talent in the country. Expanded activity will remain inhibited by the high cost of launching payloads.

1-6 Maintaining Assured Access to Space

Underlying the ability to serve any specific mission with a particular launch vehicle is the broader issue of preserving and nurturing assured access to space by the U.S. Government. The arguments for doing this would be slim if only commercial launches were involved, but military and civil programs by themselves provide adequate justification for maintaining a robust rocket industry.

Achievement and preservation of U.S. leadership in space was first mandated by Congress in the National Aeronautics and Space Act of 1958. The policy statements since that time simply reiterated that U.S. policy in this area is unswerving. As opposed to reacting to space activities of other nations, leadership in space implies a freedom to establish goals and programs on the basis of what will best serve the national interest. Acceptance of the challenge demands assured access to space, meaning dependable launch vehicles in all categories.

There is no longer a space race. Instead, the marks of a space leader rest in seriously planning the undertaking of new ventures, earnest implementation in a manner that makes the most of available budgeting, and strict adherence to short-range and long-range objectives. Space history is littered with dead programs resulting from bad starts. A reinvigorated space program demands a more refined discipline in this respect.

The U.S. rocket industry exists primarily to serve military and civil needs. Commercial rocketry in the U.S. probably could not survive by itself in the face of foreign competition by Europe's Ariane, Russia's Proton and Zenit, China's Long March II and Japan's H-2. With or without commercial rocketry, the requirement for assured access to space holds fast. More specifically, the requirement is for more than one type of vehicle, to eliminate the threat of stand-down in the event of a launch vehicle failure. The current fleet of launch vehicles satisfies this criterion. Any replacement system that may be forthcoming will have to address this issue.

Critical to assured access to space, as applied to new launch systems, is avoiding use of common subsystems across the different models. For instance, development of a single new engine that is adaptable for use in all planned launch-vehicle models would be unacceptable, since failure of the engine would result in fleet stand-down. Examination of the proposed new family of launch vehicles as first proposed in the Advanced Launch System and then in the National Launch System, shows that all have this failing if the final objective of replacing the existing fleet is met. An implied reliance on projected higher reliability of components and subsystems may be too risky for the nation's military and civil space programs.

Europe's only rocket, the French Ariane, represents a philosophy that accepts potential stand-down as a way of life. When Ariane encounters a problem, launch activities simply cease until the problem is fixed. This mode of operation is unacceptable for the U.S., where military missions take precedence and are not amenable to protracted flight delays.

Over the past few years we have seen approaches by various parties to the Congress and the National Space Council, recommending that the nation undertake development of a modern launch system, but for the wrong reasons. Reasons for developing the next generation of launch vehicles are as sound as those for developing the next generation super computer, the next generation passenger aircraft, or the next generation communications spacecraft. A part of ongoing technology is the creation of modern rocket systems that can keep pace with the future, by launching payloads efficiently, and at low cost. The Model T's have been designed and built. Now all competing nations have Model T's. Who will mainline the next model?

2

Atlas

2-1 - Background

The Atlas launch vehicle was manufactured and launched by General Dynamics Space Systems Division in San Diego, California. The company was purchased in 1994 by Martin Marietta Company. Manufacturing operations are located in San Diego and Harlingen, Texas. Final assembly and operations are carried out at the Cape Canaveral space launch facilities and at Vandenberg Air Force Base in California.

Atlas had its origin in the early U.S. ICBM program of the 1950's. It was the first of the ICBM's to reach operational status. It was also the first of the ICBM systems to be retired from service, as the DoD progressed to solid-propellant rocket systems that were better suited to operational requirements.

Atlas has played a major role in the U.S. space program since its first orbiting mission, Project SCORE, in December of 1958. Over the ensuing 35 years Atlas has successfully launched over two hundred satellites. Its demonstrated reliability over the entire history of space launches is 86 per cent. Over its last one hundred flights it has demonstrated an 88 per cent success rate. During its 35 year life, the system has undergone a series of product improvements to enhance its payload carrying capability and take advantage of modern equipment, methods, and processes. Historic Atlas events include the first U.S. manned space flight, forerunner of the Mercury series of manned flights, and the Surveyor and Mariner series of lunar and planetary explorers.

The system has evolved through many configurations in a process that has progressively improved its performance. Atlas/Able, available in 1960, was capable of launching a satellite weighing 500 pounds into geosynchronous transfer orbit. The most advanced current configuration, Atlas IIAS, will be capable of launching approximately 8,000 pounds to the same orbit. The current flight configuration, Atlas II, constitutes the basic booster that had its origin in the ICBM program, integrated with the high-energy Centaur upper stage. Centaur is the nation's only operational hydrogen-fueled upper stage, and has served as upper stage for both Atlas and Titan III. Atlas II launches take place at Complex 36A and 36B at the Cape Canaveral Air Force launch facilities.

Atlas users are the Department of Defense, NASA, and commercial organizations. Missions for the DoD are primarily communications satellites. Missions for which NASA procures launch vehicles vary, and include weather satellites operated by NOAA. Commercial missions are communications satellites presently located in geosynchronous orbit.

Atlas with Intelsat spacecraft at Florida Launch Complex 36 (General Dynamics)

Air view of Atlas launch sites 36A and 36B at Cape Canaveral (General Dynamics)

2-2 - Atlas Booster Design

Atlas is unique among launch vehicles in that its principal structure, the propellant tank, is constructed of thin-wall corrosion resistant stainless steel. The oxygen tank is forward, separated from the fuel tank by a common bulkhead. The tanks acquire structural rigidity by internal pressure, much like a pneumatic tire. Though extremely efficient, the concept has never received enthusiastic support by users, in part due to susceptibility to damage during handling or through human error. But over 500 Atlas flights have vindicated the design and Atlas remains the most efficient launch vehicle ever built.

The Atlas booster is built in two segments; the booster section and the sustainer section. The booster section contains two engines along with stored helium for pressurizing the propellant tanks during the period when the booster engines operate. The sustainer engine contains one engine and the propellant tanks that provide propellants to the engines. All engines operate at liftoff, but the boost engines cease firing between two and three minutes of flight. The boost section is then jettisoned. The sustainer section continues to operate until the part of the flight to be fulfilled by the Atlas booster has been completed. This "stage-and-a-half" arrangement, made possible by use of lightweight tanks, is performance competitive with full two-stage vehicles.

Atlas booster propulsion is provided by the MA-5 engine system built by the Rocketdyne Division of North American Aviation. Elements of the system are the boost engine, consisting of a twin turbopump assembly and two thrust chambers, and a separate sustainer engine.

An external equipment pod of the propellant tank contains Atlas components, including flight termination equipment, data acquisition, pneumatics control, and instrumentation. Flight control of the booster is provided by the Centaur avionics system, which, in addition to guidance, telemetry and navigation, supplies the vehicle sequencing signals. Steering is accomplished by means of hydraulic actuators, which swivel the engines. Roll control during sustainer engine operation is provided by separate hydrazine thrusters.

One model of the Atlas booster, Atlas IIAS, has thrust augmentation provided by four Castor IVA solid rocket boosters, which are attached to the propellant tank and boost section structure. This model, created to meet performance requirements of large spacecraft, is under development.

The Castor IVA rocket, built by Thiokol Corporation, is 37 feet long and forty inches in diameter. During launch, two motors are ignited at lift-off and two are ignited during flight.

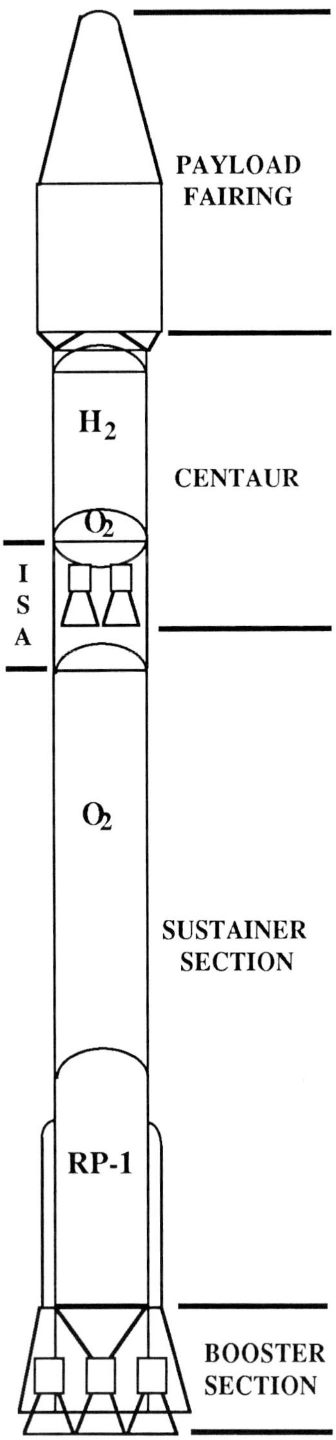

Major components of the Atlas Launch Vehicle

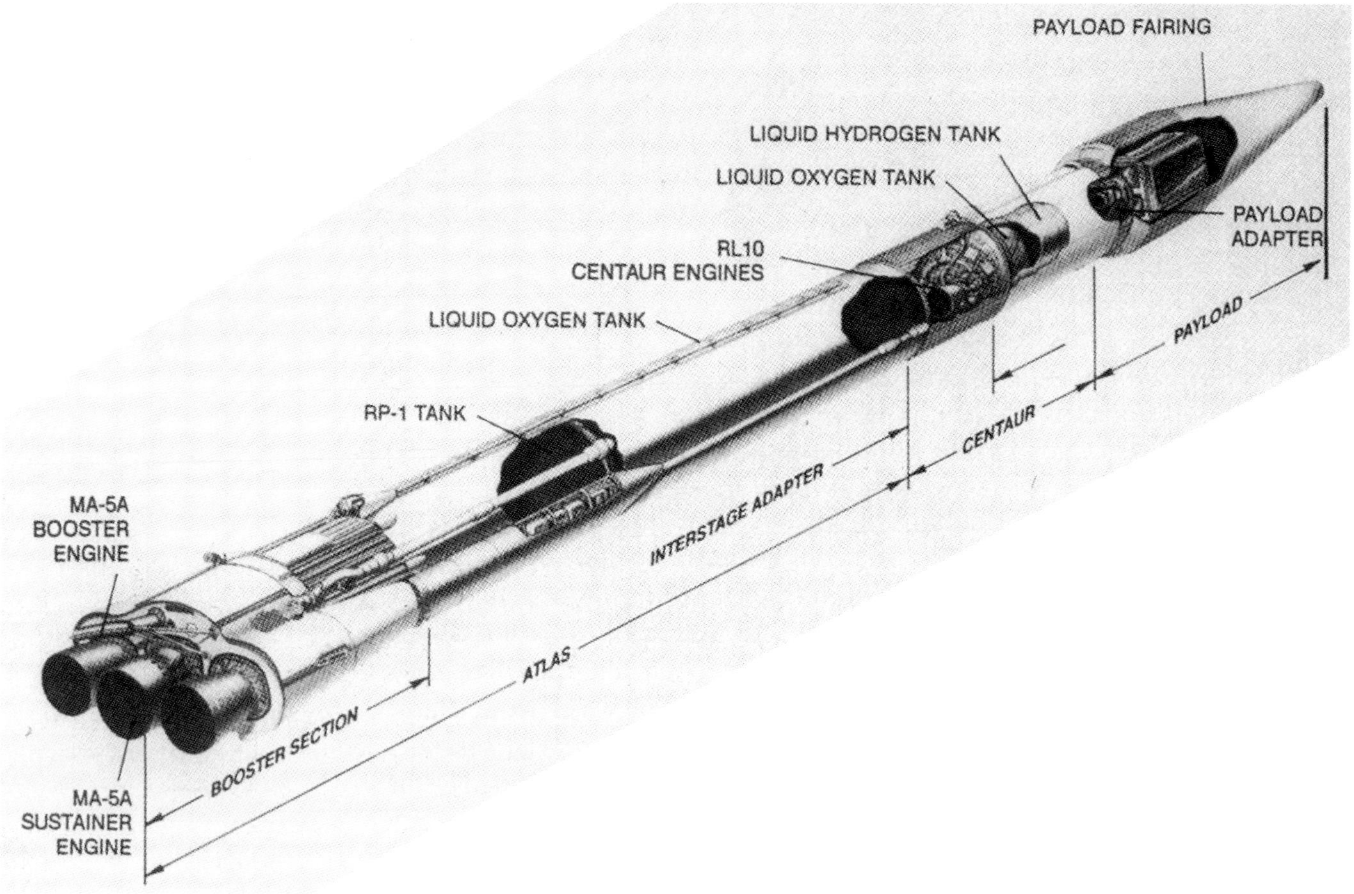

Artist's rendition of the Atlas II with its payload (General Dynamics)

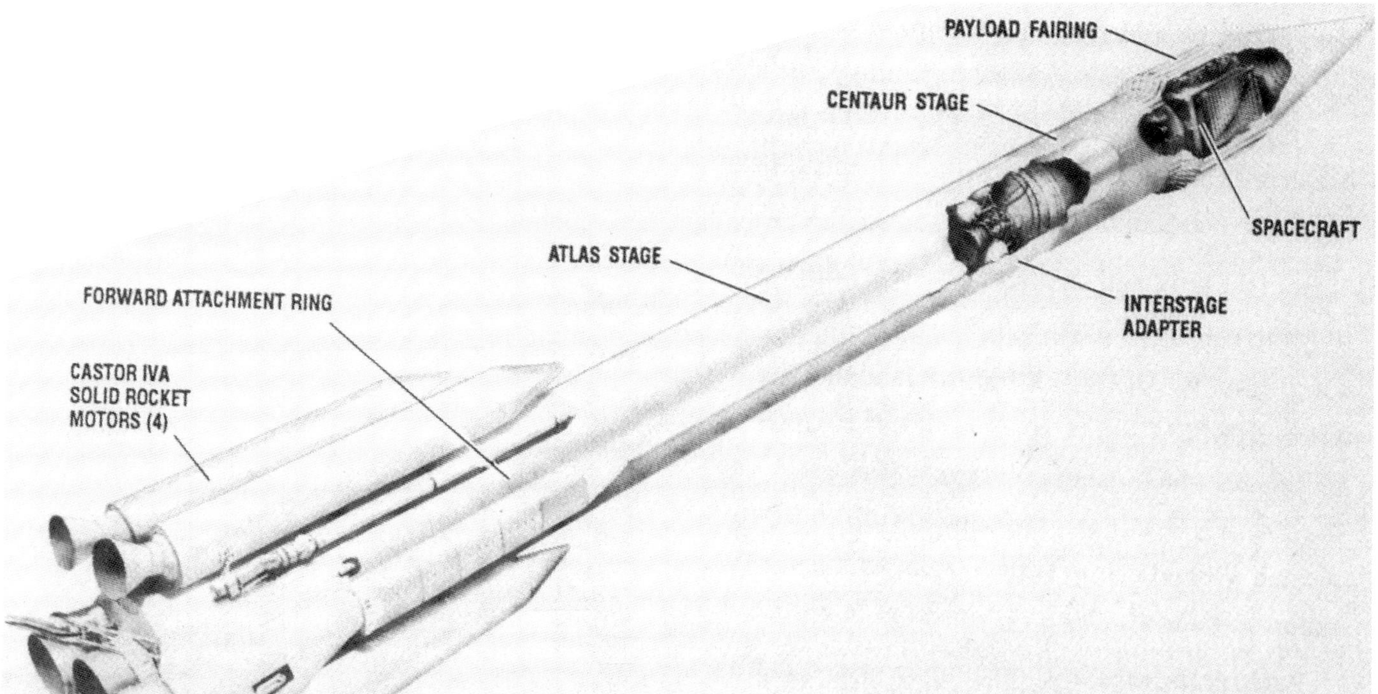

Artist's rendition of the Atlas IIAS showing solid rocket boost augmentation (General Dynamics)

2-3 - Centaur Upper Stage

Centaur is the nation's first liquid hydrogen powered upper stage. Its development was initiated by the Air Force in the late 1950's. In the course of events, the requirement for the stage changed and responsibility for its development was turned over to NASA. Launched atop Atlas and Titan, this high-energy stage was exactly what NASA needed to perform its planned lunar and planetary exploration missions.

Centaur is structurally linked to Atlas through an interstage adapter. The booster is separated from the upper stage by a shaped-charge device located at the forward ring of the adapter.

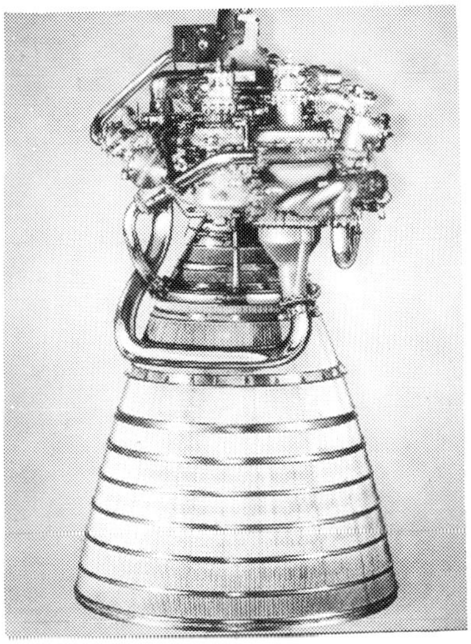

16,500 lb thrust RL10A-3-3A Centaur rocket engine (Pratt & Whitney)

The propellant tanks, like those of the Atlas booster stage, are constructed of thin-wall corrosion resistant stainless steel. Centaur propellants are liquid hydrogen and liquid oxygen. The propellant tanks form the principal structure of the stage and share a common, vacuum insulated double-wall bulkhead. The hydrogen tank is insulated to minimize boiloff losses and inhibit condensation of air on its surfaces when it sits loaded on the launcher.

A short adapter ring and conical equipment module are attached to a ring at the forward end of the cylindrical section of the hydrogen tank. The adapter provides attachment rings for securing the payload and the payload fairing. Centaur avionics components are mounted on the equipment module. These provide guidance, communications, and power and control functions for Atlas and Centaur.

Centaur is powered by two Pratt & Whitney RL-10 engines. These engines are available in several models, a result of the engine company's upgrade efforts directed toward improved stage performance. Early models of the engine were rated at 15,000 pounds thrust. Up-rated models of the engine are rated as high as 20,800 pounds thrust.

20,800 lb thrust Centaur engine with extendible nozzle (Pratt & Whitney)

Centaur has been used as an upper stage for Atlas and Titan launch vehicles. The Titan launches carried NASA's highly successful Viking and Voyager missions. During the 1980's a design and development effort was initiated by NASA to integrate Centaur with the Space Shuttle. The initial objective was to provide propulsion for the Galileo mission to Jupiter. Shuttle Centaur had fifty per cent more propellant capacity than the Centaur used with Atlas. With its hydrogen tank increased in diameter to 14 feet from its original 10 feet, it was called the "wide body" Centaur. However, a growing concern regarding safety for the astronauts, magnified considerably by the Challenger accident, led to cancellation of this effort. Several hundred million dollars were spent on this program before its cancellation. Chapter 14 includes more information on Centaur and its future derivatives.

2-4 - *Atlas Capability*

The Atlas launch vehicle family consists of four models. They are characterized by incremental improvement in performance, brought about by user requirements for bigger and heavier satellites. Some performance improvements were achieved through vehicle weight reductions. Most however came from increases in engine thrust, which allowed designers to increase propellant capacity in both Atlas and Centaur.

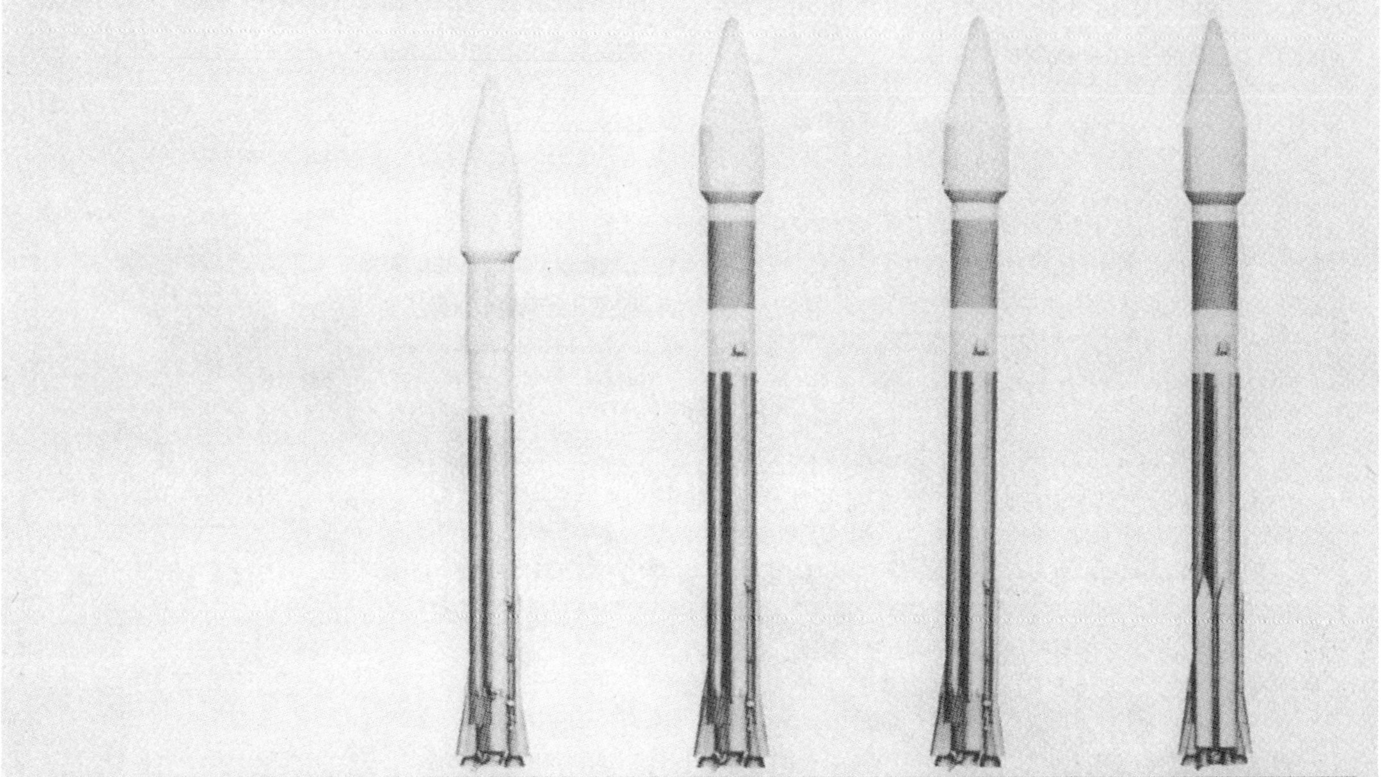

	ATLAS I	ATLAS II	ATLAS IIA*	ATLAS IIAS*
LARGE FAIRING	4,970 LB (2255 KG)	6,200 LB (2810 KG)	6,710 LB (3045 KG)	8,150 LB (3700 KG)
MEDIUM FAIRING	5,240 LB (2380 KG)	6,500 LB (2950 KG)	6,970 LB (3160 KG)	8,450 LB (3830 KG)
INITIAL LAUNCH CAPABILITY	JULY 1990	DECEMBER 1991	JUNE 1992	1993

*BLOCK 1 PERFORMANCE LEVELS

PAYLOAD SYSTEM WEIGHT CAPABILITY TO GTO AT 27.0-DEGREE INCLINATION. PSW INCLUDES SPACE VEHICLE, ADAPTER, & MISSION-PECULIAR HARDWARE. TYPICAL ADAPTER AND MISSION-PECULIAR HARDWARE IS 150 LB (68 KG)

The Atlas launch vehicle family (General Dynamics)

Atlas II A reached the practical limits of extending the performance capability of the basic launch vehicle. Further improvement is to be achieved by adding solid rocket "strap-on" boosters to the Atlas. The designation for this model is Atlas IIAS.

Atlas I is primarily devoted to commercial launch services. Atlas II serves DoD, NASA requirements as well as commercial missions.

Payload accommodations provide a choice of two sizes of fairings. The medium fairing is a skin-stringer, two piece aluminum structure. The fairing and associated support structures provide a protective enclosure for Centaur avionics, as well as supporting and protecting the payload while on the launch pad and during flight through the atmosphere.

The large fairing widens the diameter from eleven feet for the medium fairing to fourteen feet. Flights with the large fairing carry a small performance penalty due to its added weight.

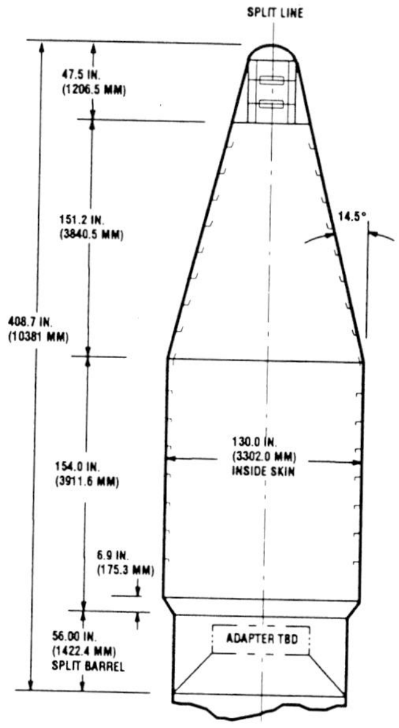

Medium payload fairing
(Gen Dynamics)

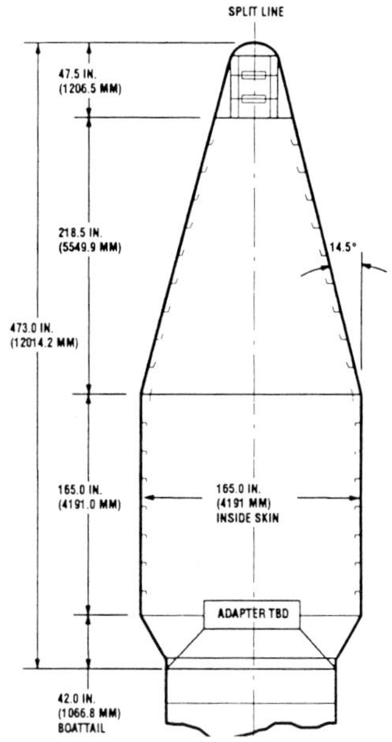

Large payload fairing
(General Dynamics)

Atlas fairing during installation (Gen Dynamics)

2-5 Growth Capability

General Dynamics Space Systems Division maintains a modest program of continuous improvement to the Atlas launch vehicle, aimed at gains in both performance and flight reliability. These improvments are in a large part dependent on government funding. Among the advancements are a single-engine Centaur, substitution of aluminum-lithium alloys for aluminum to obtain lighter weight structures, avionics advancements and electromechanical actuators for engine gimbaling.

Models other than the Atlas II and Atlas IIAS have been designed, but not produced. These include Atlas JII and Atlas JIIS. Neither of these vehicles uses the Centaur upper stage, their missions typically being to launch payloads into low-earth orbit. The JIIS uses solid rocket motor boost augmentation. These vehicles have more propellant capacity, in an amount roughly equal to the weight of the Centaur stage.

Performance improvements to Atlas II AS can be obtained by increasing the number and size of the solid rocket boosters and increasing Centaur propellant capacity.

An attractive method of improving Atlas performance involves use of liquid propellant strap-on boosters which utilize the same engines as the basic booster. These rockets can be constructed at low cost, using propellant tanks that are built to standard ASME code, which can be fabricated by certified weld shops. Unit cost of engines is reduced by increasing the production base. A significant advantage of the arrangement is that the Atlas structure is relieved of the high bending loads normally experienced in flight, resulting in a far more robust arrangement than currently exists. With this improvement, Atlas II performance would be raised from 8000 pounds to geosynchronous transfer orbit to over ten thousand pounds.

Reiterating a subject of the first chapter, very little occurs in the rocket industry beyond the preliminary design and research phase. Improvements come about only if a mission requirement appears. At that point, whatever improvements are appropriate to the mission are offered to the prospective customer in the company's business proposal. Competing companies may or may not do the same thing, depending upon whether their product already has the capability to perform the mission.

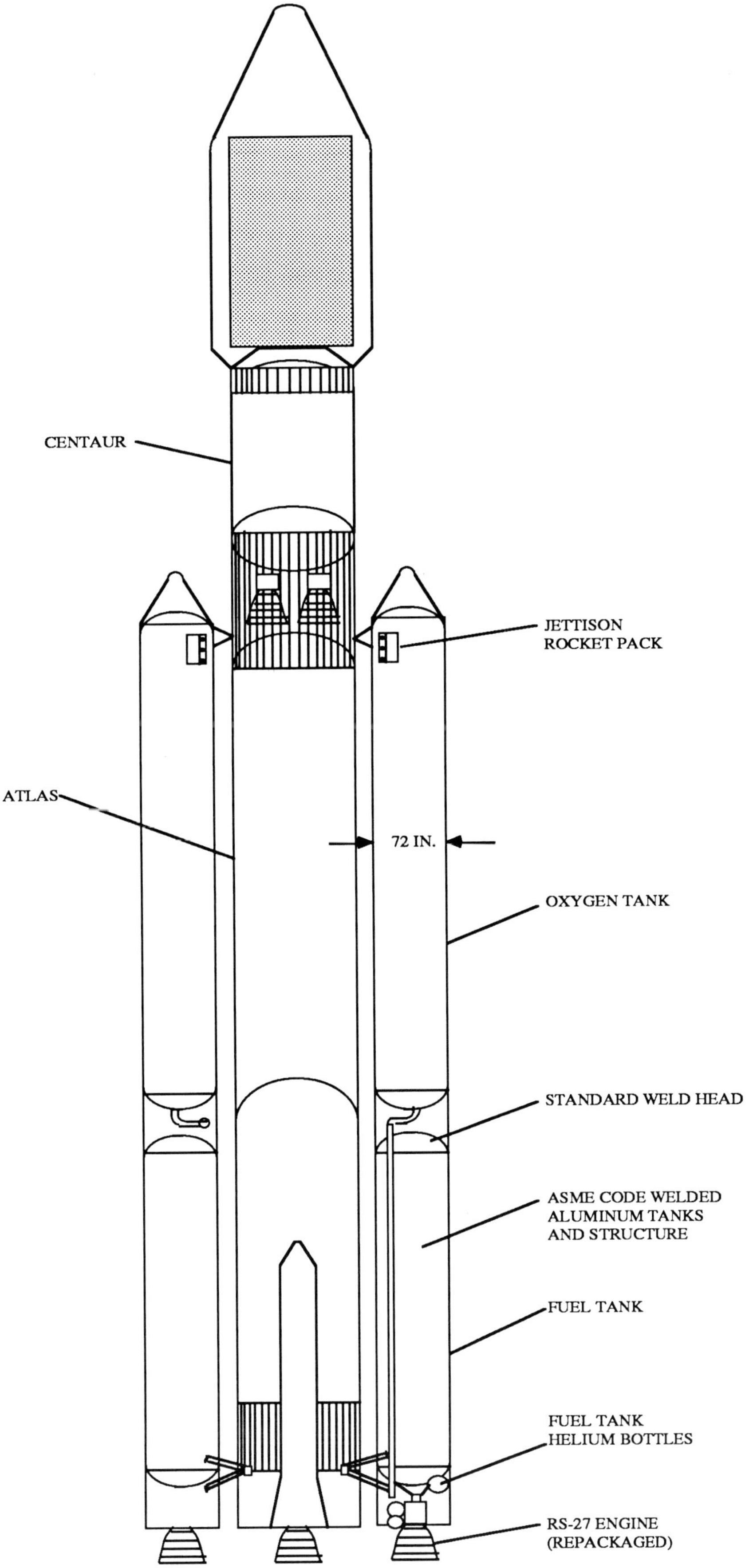

Atlas with liquid propellant strap-on boosters for heavy payload missions.

2-6 Outlook

The future of Atlas is somewhat clouded by events brought about by the end of the Cold War. A fall-off in the defense industry has produced different responses by the major industrial firms that for over forty years have been engaged in developing and producing new, advanced weaponry. General Dynamics has elected to downsize, perhaps leading to complete dissolution, as evidenced by a sell-off of most of its operating divisions. This one time "Strong Company for a Strong Country" is now but a shell of its former self, possessing only the Electric Boat (submarines), Land Systems (tanks) and Space Systems Divisions (space launch vehicles) and a few minor holdings.

In April of 1994 an announcement was made of the sale of the Space Systems Division to Martin Marietta Company. By this act, the number of major producers of launch vehicles in the U.S. has been reduced by one third. In theory, the merger will reduce manpower requirements and thus result in more economical production of launch vehicles. Martin Marietta's ultimate plans for the two vehicles produced by General Dynamics, Atlas and Centaur, are not known.

The Space Systems Division was injured by a series of flight failures during the past year, resulting in a reluctance by payload producers to use the system for launching their spacecraft. Two Centaur failures, similar in nature, followed by an Atlas booster failure, occurred during the past year. But fixes have been prescribed and the Atlas program continues. Recent successful flights of the Atlas II and Atlas IIAS appear to demonstrate that these troubles have been surmounted. If Congress approves Air Force recommendations that for the near future the nation should upgrade its existing launch systems, the Atlas system stands to benefit greatly by programs designed to improve its reliability. Significant funding has already been provided to upgrade the Atlas system.

Atlas II is the only U.S. launch system in production able to meet payload lift requirements between the capabilities of Delta II and Titan IV. As such, it seems assured of a future market, if only to supply Air Force and NASA requirements. The vehicle faces severe competition by foreign countries for commercial missions, where lower prices and better reliability are promoted. Operating in the commercial launch arena will call for more aggressive cost control than Martin Marietta has been accustomed to in its predominately military business base.

The Air Force is moving ahead with plans to install Atlas II launch capability at Vandenberg Air Force Base in California. This will be the first time that hydrogen-fueled upper stages have been flown from the west coast. When this installation is completed, there will be a total of three launch platforms for Atlas launches, two east coast and one west coast.

3

Delta

3-1 Delta's Origins

Like Atlas and Titan, the Delta launch vehicle originated in the early ballistic missile programs that were undertaken by the Air Force in the 1950's. Delta's ancestor was the Thor intermediate range ballistic missile developed by Douglas Aircraft. The IRBM's were installed at sites in Europe, but were retired after only a few years.

In 1959, NASA-Goddard Space Flight Center contracted with Douglas Aircraft (now McDonnell Douglas Corporation), to produce twelve space launch vehicles, using components from the Thor and Vanguard programs. This was the beginning of the Delta program. The original vehicles were named Thor Delta, Delta being the name assigned to a liquid propellant upper stage created for this launch system. For many years Goddard continued as the procuring agency for the McDonnell Douglas vehicle, supplying launch services for military, civil, and commercial missions.

Eventually the Thor name was dropped. McDonnell Douglas stopped making the Delta upper stage, replacing it with a new device known as the Second Stage Propulsion System (SSPS). The complete launch vehicle came to be identified as Delta. The third stage of Delta is one of several STAR solid rocket motors built by Thiokol Corporation.

Over time, Delta has undergone many modifications to improve its payload carrying capability. Upgrades in engine thrust, addition of solid rocket "strap-on" boosters, and improvements in the upper stage were some of the major changes that were adopted.

Like Atlas, Delta was commercialized after the Shuttle Challenger accident. It has a continuing market for military and civil missions. Atlas and Delta compete for the same missions. The determining factor for who wins is usually the payload weight, with the heavier payloads going to the higher performing Atlas.

Delta is presently the lowest cost of the western launch vehicles, as represented by Atlas, Titan, Delta, and Europe's Ariane. It is the most reliable launch vehicle, having been successful in 97 of its last one hundred flights, and demonstrating 100 % success in the last forty or so missions. Only the Space Shuttle comes close to matching this record, with one failure in 53 flights.

Delta has launch sites at both east and west coast facilities.

3-2 Delta II System Description

The latest versions of Delta were begun after the Challenger accident to support an Air Force decision to off-load the Global Positioning Satellite (GPS) from the Shuttle manifest and accelerate its availability. The Delta II designation coincided with award of the contract for Medium Launch Vehicle I to McDonnell Douglas. Two versions of Delta II were created, but only the higher performing one is now produced.

The first stage of Delta uses liquid oxygen and RP-1 propellants and is powered by a single RS-27 engine produced by Rocketdyne. The RS-27 Rocket engine has a sea-level thrust of 207,000 pounds. This engine has compiled one of the most consistent records in the history of rocketry, with a success rate of 100%. The RS-27 A is the latest version of the Delta propulsion system. Modifications to the RS-27 include an increased capability heat exchanger, an increased capacity lubrication system, and an increase in thrust chamber expansion ratio from 8:1 to 12:1. This modification produced a specific impulse increase of seven seconds over the RS-27 model. The RS-27A is now the main propulsion system for the Delta II family of launch vehicles.

Two low-thrust Rocketdyne vernier engines provide roll control during main engine burn, and attitude control during the short period after main engine cutoff and second stage ignition. The propellant tank structure is equipped with attachment points for solid propellant strap-on rocket motors, which provide thrust augmentation during boost phase. Nine solid rocket motors are attached to the core stage. Six of these motors ignite at liftoff and the remaining three are ignited during flight. The solid motors, designated GEM, are manufactured by Hercules Aerospace Company.

The two eight-foot diameter propellant tanks are constructed of aluminum alloy. The cylindrical shells are of machined "isogrid" construction for structural strength. The tanks are connected by a cylindrical adapter that houses control electronics, ordnance sequencing equipment, and telemetry.

Delta's second stage, the SSPS, uses nitrogen tetroxide and Aerozene 50 for propellants. These propellants ignite spontaneously when combined, so the engine does not require an ignition system. The SSPS is powered by an Aerojet engine designated the AJ10-118 K. This engine is an advanced version of an engine developed by Aerojet for a Titan III upper stage called Transtage, and for the second stage of the Japanese N-II rocket. The stage has no turbopump, using instead gaseous helium to pressurize the tanks to high pressure. Guidance equipment for the first and second stages is mounted forward of the propellant tanks on the second stage.

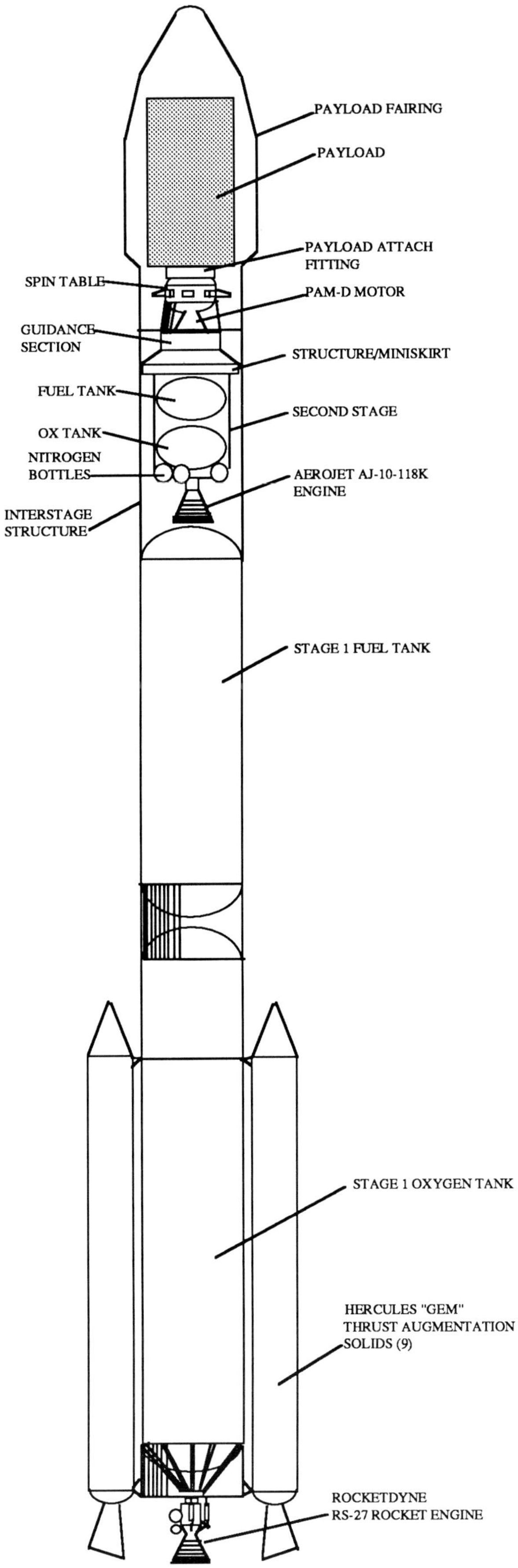

Delta II main components

The third stage uses the STAR-48B (PAM-D) solid rocket motor built by Thiokol Corporation. The rocket is mounted to a spin table that is attached to the forward ring of Delta's second stage. This motor weighs approximately 4700 pounds and has a thrust level of 15,100 pounds. Attached to its forward end is the payload attach fitting, to which the satellite is secured. Third stage stabilization is achieved by spinning the solid motor with its satellite before separation from the second stage.

Delta II operates in a two stage version for low-earth orbit missions. Four standard payload fairings are provided, two at 9.5 feet in diameter and two at ten feet in diameter.

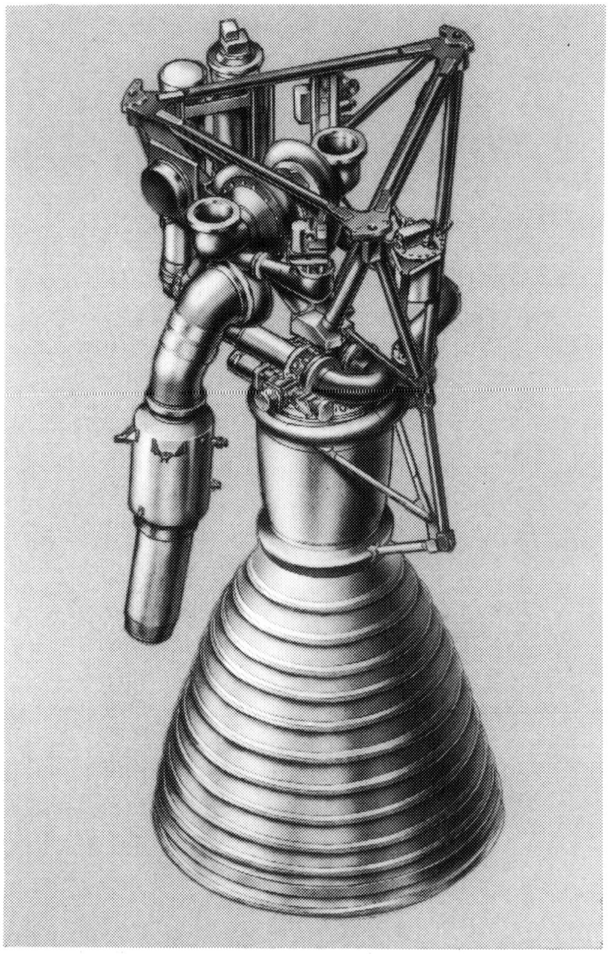

Rocketdyne RS-27 and RS-27A rocket engines used on the Delta and Delta II launch vehicles respectively (McDonnell Douglas)

Initial Delta II commercial flight carrying Nato IVA, the first of the NATO satellites (McDonnell Douglas)

3-3 Delta II Improvements

Delta's position in the market provides ample reason for McDonnell Douglas to pursue a program of product improvement. Such activity ensures a hold on its present market and opens opportunities to broaden the system's range of applications.

Most of the company's upgrading activity is concentrated in two areas: installation of modern avionics systems and improving the vehicle's range safety system.

Launch vehicle manufacturers are driven to constantly improve their avionics systems simply because of the rapidly changing nature of the electronics industry. What may have been an up-to-date computer in 1980 is quite likely to be out of production by 1990 due to obsoletion and scarcity of parts. Continued production of old equipment designs can be prohibitively expensive, especially for the limited production runs common to launch vehicles, whereas it makes good sense to take advantage of new innovations that actually lower the cost.

Downsizing of electronic equipment has also made possible introduction of multiple redundant circuits that serve to improve vehicle reliability at no cost in added weight.

The fully redundant avionics system that McDonnell Douglas and other manufacturers are adopting has three complete command paths, so that loss of any one path does not affect the mission. Single failure tolerance characterizes the next generation of launch vehicle avionics. In contrast, Man-rated vehicles require double-failure capability. The record of launch failures due to avionics does not seem to justify the added cost and complication of double-failure tolerant avionics systems for unmanned vehicles at this time.

A large part of the design effort toward achieving a single failure system centers on efficient packaging. Design goals aim to install all critically related parts in a single box rather than three or more separate boxes. This drives the cost down and has a side effect of reducing power requirements so the vehicle requires fewer batteries. The new Delta box contains six Allied Signal laser gyroscopes, six Sundstrand accelerometers, and three Allied Signal computers.

Range safety equipment requirements have become much more stringent since the Challenger accident. McDonnell Douglas and other launch vehicle manufacturers are consequently obligated to invest in system improvements. Delta adopted new avionics that are less susceptible to interference. The new design features three separate ordnance signal devices that are independent of the rest of the vehicle's avionics.

Growth of Delta II to greater payload carrying capability can be obtained by adding bigger booster rockets, increasing the thrust of the main engine and "stretching" the propellant tanks, adding more powerful upper stages, or combinations of these methods. The most significant gains can be had by improving the second and third stages.

In 1991, the Air Force began work on the Solar Electric Orbital Transfer Vehicle (SEOTV) as a method of moving the Global Positioning Satellites into their operating orbits. A forerunner to this device is the Electric Insertion Transfer Stage Experiment under development by TRW, which would later evolve into an operational transfer stage. McDonnell Douglas and TRW are studying the future transfer stage. The SEOTV would be Delta's third stage, replacing the solid rocket motors presently in use.

In theory, electric propulsion has the potential of doubling the performance of Delta, but this gain may be heavily penalized by the weight of the electric stage once all the operational requirements are known. The problem centers on providing the electrical power for generating thrust. In the case of the SEOTV, large solar panels would be required, as well as heavy power conditioning equipment. Electric powered transfer vehicles are poorly regarded by the commercial world because of the excessively long transfer time before the satellite is placed into its operating orbit. Weeks and months lost during transfer translate directly into lost revenue, which is not acceptable. The method of transfer is acceptable when the user is amenable to long transfer times, which may be the case for some DoD missions. Due to budget problems, the SEOTV is currently in limbo, so the future of electric propulsion remains quite distant.

Delta II's performance could be greatly upgraded were McDonnell Douglas to develop a hydrogen-fueled upper stage similar to, but smaller than Centaur. The company has toyed with the idea since the 1960's, having then configured the "HOSS" upper stage, a single engine, eight foot diameter, aluminum bodied vehicle that sat atop the boost rocket.

The Post-Boost Vehicle that is used on the MX missile has been promoted by Rocketdyne as a future upper stage on Delta. This device could provide spacecraft maneuverability, multiple burns, and long orbital stay time, since it uses earth storable propellants instead of volatile cryogenics. But the Post-Boost Vehicle is currently out of production. About a dozen residual stages were transferred to NASA by the Air Force for possible future use in space propulsion. Resumption of production to meet Delta's needs would be costly.

3-4 Outlook

McDonnell Douglas competed against General Dynamics and won an Air Force contract for Medium Launch Vehicle III in 1993. The Delta vehicle is designated MLV III and is slated to launch medium weight DoD payloads from east and west coast launch sites. This contract assures production of Delta II well into the next decade.

Delta appears to have an assured market for launches within its payload range for both civil and commercial missions. The current trend toward smaller and lighter weight spacecraft design works favorably for McDonnell Douglas, helping it to wrest payloads from competitors with greater launch capability. New types of communications systems, notably mobile communications, are being designed in a range of sizes that fit Delta capability.

A dark cloud on the horizon for McDonnell Douglas is the emerging Lockheed launch vehicle family designated LLV1, LLV2, and LLV3. The Lockheed system can significantly undercut Delta's price for future launchings. As builder of the satellites for Motorola's Iridium mobile telephone system; as partner with Russia's Khrunichev organization which will provide Proton launch vehicles to launch some of the Iridium satellites; and as producer of a new line of vehicles targeted toward small satellites, Lockheed is positioning to challenge the entrenched manufacturers of launch vehicles on their own ground.

Like General Dynamics (now Martin Marietta) with its Atlas II vehicle, McDonnell Douglas keeps an array of performance upgrade solutions on the back burner, awaiting new customer requirements. The company is slow to risk capital to develop launch capability in advance even though such efforts would allow Delta to compete with Atlas, Titan II, and Europe's Ariane in their class of payloads. A conservative management is reluctant to venture outside existing boundaries set by Delta's performance. Absence of action on a hydrogen-fueled upper stage, in spite of its clear advantages, illustrates this ingrained restraint.

It can be argued that McDonnell Douglas foresees the end of its Delta launch vehicle in light of its ongoing efforts to develop the Delta Clipper, a fully reusable single-stage-to-orbit rocket. But Delta Clipper, or any new concept for a reusable launch vehicle, faces a long road before the nation is ready for this leap of faith. The Space Shuttle has effectively destroyed any notion that a reusable launch vehicle will be more economical than an expendable one. The foundations for this reasoning have to be rebuilt, and as yet no one has any idea of how to go about it.

McDonnell Douglas presently produces the most economical and reliable launch vehicle in the western world. That fact alone assures Delta's active participation in space launches for at least another decade.

4

Titan

4-1 Background

Titan launch vehicles are manufactured by Martin Marietta Company, Astronautics Group, in Denver, Colorado. The Titan family consists of Titan II, Titan III and Titan IV. Titan II's are former ICBM's which are undergoing conversion to space service. Titan III is a space launcher derived from an earlier space version of the ICBM and served the Air Force and NASA successfully for many years, in several models. It was replaced by Titan IV during the latter half of the 1980's.

Titan IV is the nation's largest unmanned launch vehicle. Its development was initiated in 1984 when the Air Force determined that the nation must have a backup for the Space Shuttle. It was then known as the Complementary Expendable Launch Vehicle (CELV) and was intended to be activated in the event that the Shuttle was unable to meet its commitments. Four years were allowed for its development, including modifications to launch facilities at Cape Canaveral Air Force Station.

The Space Shuttle Challenger accident in January of 1986 resulted in national space policy changes to reinstate expendable launch vehicles as the primary means of launching automated spacecraft. Following the accident the CELV title was dropped and Titan IV became the principal carrier of critical, high mass payloads. Orders for delivery of Titan IV's were upped from an initial 10 vehicles to a total of 41 to accommodate a backlog of payloads that were removed from the Shuttle manifests. At the same time plans were expanded to create a Titan IV launch capability at Vandenberg Air Force Base in California.

NASA decided at this time to abandon development of the Space Shuttle version of Centaur. But this did not eliminate requirements for its use by the Air Force. As part of the restructuring, the Air Force/Martin Marietta team assumed sole responsibility and successfully completed development of the wide-body Centaur for Titan IV.

Titan space launch vehicles have an overall success rate of 93 per cent, covering a total of 173 flights. The success rate for the last 100 flights is 94 per cent. This system, like Atlas, does not appear to have responded to attempts to improve reliability.

As of mid-1994, Titan IV has experienced nine flights. The seventh flight failed as it neared termination of boost phase propulsion. The eighth and ninth flights were successful, and included the wide-body Centaur as the upper stage.

Titan IV launch at Cape Canaveral Air Force Station in Florida (Martin Marietta)

4-2 Titan IV Launch System

Titan IV consists of a core vehicle with two solid propellant "strap-on" booster rockets. It operates with or without upper stages, depending on mission requirements. The core vehicle consists of two separate stages, measuring ten feet in diameter. The propellant tanks and structure are made of aluminum alloy. The stages are powered by engines produced by Aerojet Corporation of Sacramento, California. Thrust of the boost engines in a vacuum environment is 548 K pounds. Stage II is powered by a single engine, which has a vacuum thrust of 105K pounds.

Core vehicle propellants are nitrogen tetroxide and Aerozene 50, which is a hydrazine based fuel. When the Titan II ICBM was designed, the Air Force required a system that was ready for launch, meaning that missile tanks were kept fully loaded with propellants while on station in the silos. Cryogenic propellants were therefore ruled out. The selected propellants, though highly toxic, remain with the system to this day.

Titan IV has a gross mass of 1.9 million pounds and a liftoff thrust of 3.2 million pounds.

Titan IV core vehicle rocket engines (Aerojet)

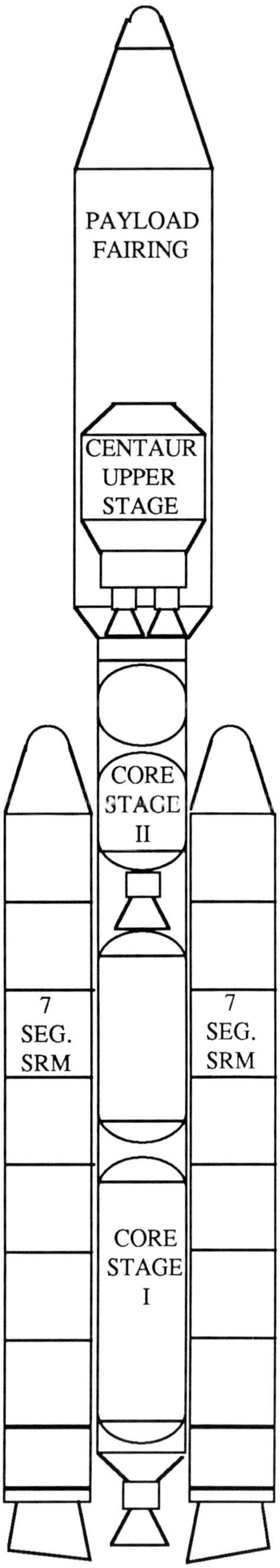

Titan IV major components

Steering the core vehicle after the solid rocket motors have been jettisoned is accomplished by gimbaling the liquid-fueled engines. Hydraulic actuators are used for this purpose.

Titan IV carries its own avionics equipment to enable operation when an orbit transfer stage is not present. A transfer stage carried atop Titan IV is equipped with its own avionics system.

The solid rocket motors (SRM's) presently in use on Titan IV are built by United Technologies. These motors measure 10.2 feet in diameter by 112 feet in length and have a thrust of 1.6 million pounds in a vacuum environment. Due to their great weight, the SRM's are shipped in segments and assembled at the space-launch complexes. Roll control and steering of the vehicle during boost phase is accomplished by injecting nitrogen tetroxide laterally into the exhaust flow stream of the SRM's to shift the direction of thrust.

An advanced solid rocket motor, termed the "solid rocket motor upgrade" (SRMU) is currently undergoing development by Hercules Aerospace Company. Current Air Force plans are to replace the United Technologies seven segment motor when the new one has been qualified for flight. The Hercules motor has only three segments and will enable a substantial increase in Titan IV performance. More about the SRMU in Chapter 7.

Titan IV's are assembled in a vertical assembly building (VAB) at the launch complexes. The completed vehicle is then rolled out on tracks in a vertical position to the launch pad. The solid rocket motors are separately assembled in a solid rocket motor assembly building before being transferred to the VAB for final assembly.

Complexes 40 and 41 at Cape Canaveral Air Force Station are Titan launch sites. Complex 41 is operational and complex 40 is undergoing modifications from a Titan III configuration to a Titan IV. When complete, the Air Force will have dual capability at Cape Canaveral to launch large payloads into low-earth orbit with Titan IV alone, or high-energy missions using the IUS or Centaur upper stages.

Titan IV space launches are also operational at Vandenberg Air Force Base Complex 4 East in California. This facility provides a capability to launch large payloads into high inclination orbits, which cannot be done in Florida because of over-fly of populated areas.

Titan IV is in reality a new launch vehicle system coming on line. It is the nation's largest expendable launch system and is designed to carry large payloads into low-earth orbit, high-earth orbit or escape missions. Its most powerful version, Titan IV/Centaur was first flown in 1994.

4-4 Titan IV Capabilities

Titan IV is available in several configurations to meet various payload requirements. The principal differences reside in presence or absence of orbit transfer stages, and size of payload fairings. Transfer stages are used for geosynchronous orbit and other high-energy missions and are launched from Cape Canaveral Air Force Station. Upper stages in current Titan IV designs are Centaur and IUS. Payloads without transfer stages are also launched from Cape Canaveral.

Titan IV launches from Vandenberg Air Force Base are presently confined to heavy payloads carried to high-inclination low-earth orbits.

Titan IV payload weight carrying capability will increase with introduction of the SMRU's around the turn of the century.

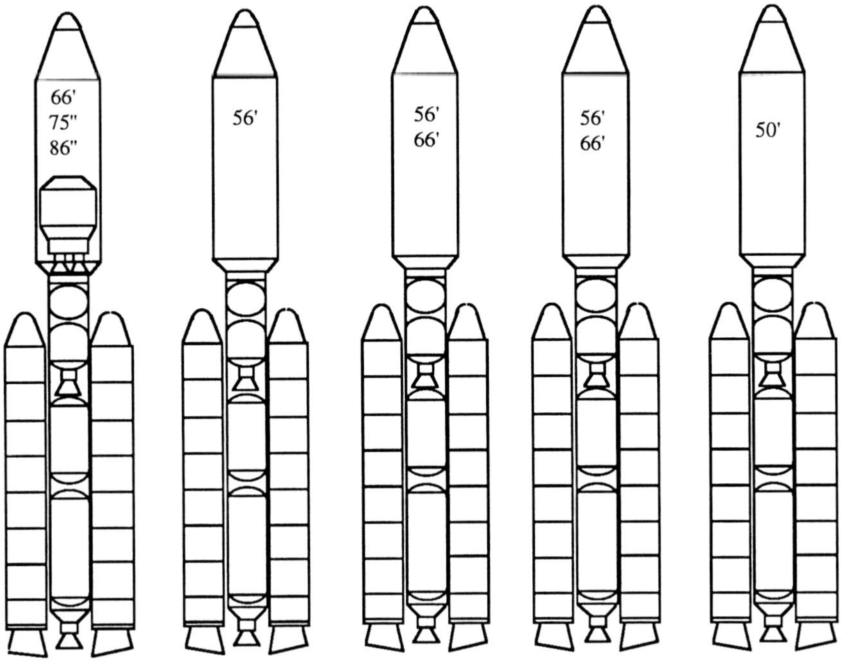

UPPER STAGE	CENTAUR	IUS	NONE	NONE	NONE
LAUNCH SITE	CCAFS	CCAFS	CCAFS	VAFB	VAFB
PAYLOAD (LB)	10,000 (GEO)	8000 (GEO)	38,780 (LEO)	31,100 (LEO)	29,600 (LEO)
PAYLOAD (SRMU)	12,700 (GEO)	9000 (GEO)	47,800 (LEO)	38,800 (LEO)	36,700 (LEO)

Summary of payload lift capabilities of the Titan IV family of launch vehicles (Martin Marietta)

4-4 Titan II

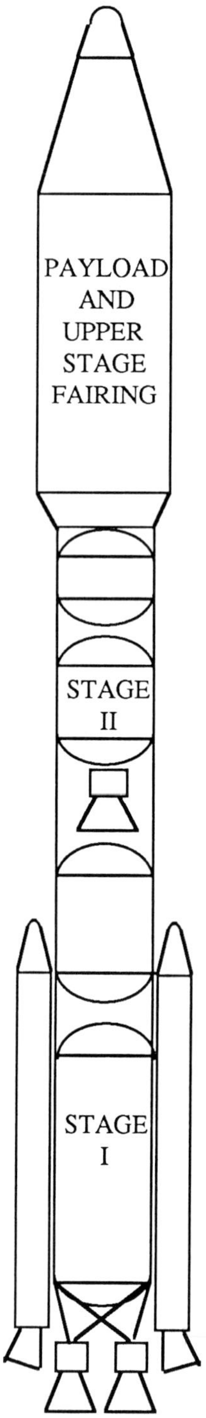

Titan II space launch vehicle shown with solid rocket strap-on booster option.

When the Titan II ICBM fleet was dismantled, more than fifty Titan II's were made available for use as space launchers. The idea had already been put into practice in the past, when nearly all retired Atlas ICBM's were put to the same use in a highly successful and economical program instituted by the Air Force and General Dynamics at Vandenberg Air Force Base.

The retired missiles were placed in storage, to be released as needed for future programs. Space use of obsoleted ICBM's makes good economic sense, since the vehicles can be reconditioned and converted to flight status for a few million dollars. Production costs for used Titan II's range from $15M at production rates of ten per year to $31M at one per year. Conversion tasks include refurbishment of the hardware to an as-new state as well as incorporation of upgrades in several subsystems. Launch costs are approximately $4M per flight.

A number of reliability, performance, and operability improvements to Titan II are underway, including multiple burn for the second stage, fault tolerant guidance, lighter and more economical structures and Stage I thrust augmentation. Of these improvements, the most significant for improved performance is thrust augmentation, accomplished by adding strap-on Castor IV/GEM class solid rocket motors to the boost stage. This method of thrust augmentation has long been practiced by McDonnell Douglas on the Delta rocket and now by General Dynamics on its Atlas IIAS.

Titan II's are launched at Vandenberg Air Force Base. Foreseeable demand for flights at high-inclination orbits that fall within Titan II's capability is about two per year. At this rate, the life of the system could extend to the year 2020.

The respectable reliability of the Titan Stage I and Stage II foretells possible entry of Titan II into future medium payload launch vehicle competitions. An updated two-stage vehicle based on the Titan core design, equipped with the new Thiokol Castor 120 strap-on booster rockets, would be a formidable system by any definition. Since the basic improvements are already present in the Titan IV core stage, and the Castor 120 is already developed, the nonrecurring costs for this medium launch vehicle should be low.

As was the case with the Atlas ICBM's, Titan II was initially looked upon as a threat to the ongoing production of medium launch vehicles as defined by Delta II and Atlas II. The Air Force's planned prudent use of this resource over nearly three decades should alleviate this concern.

4-5 Titan III

Titan III was developed as a result of 1961 studies by DoD and NASA to examine how to perform missions that could not be accomplished by the biggest rockets then in operation , Titan II and Atlas. Titan III was the outgrowth of this concern. The underlying concept for this launch vehicle was drawn from technologies made available by the Titan and Minuteman ICBM programs. The original concept adopted the Titan II as the core stage, boosted by two solid propellant rockets. A liquid propellant third stage, called the Transtage, completed the assembly. The launch vehicle was developed rapidly. By September of 1964 the first Titan III had flown.

Titan III appeared in several versions over the next twenty years, representing consecutive upgrades in core vehicle and solid rocket motor capability, as well as different upper stages. Transtage, IUS, and Centaur were used at various times with Titan III. One version, Titan IIID, was used for low-earth orbit missions and required no upper stage. Titan III evolved to Titan IV in the 1980's. Titan IV features stretched stages I and II as well as seven-segment solid rocket booster motors.

In the mid-eighties, Martin Marietta decided to compete for launches of commercial satellites and began production of Titan III again for this purpose. The first commercial flight of Titan III failed to release its payload. After that failure the company's ambitions dimmed for Titan launch vehicles in the commercial world.

Martin Marietta contracted with NASA to launch the Mars Observer spacecraft on Titan III, which took place in late 1992. But future plans, such as the Cassini mission to Jupiter, call for use of Titan IV due to its higher performance (NASA lost communication with Mars Observer in August, 1993, as the spacecraft approached Mars).

The future of Titan III is an open question. The vehicle remains a more economical option than Titan IV. It could be a useful device in the emerging small satellite business, where as many as seven or eight satellites are launched on a single flight. The recent sharp cutback in Air Force requirements for Titan IV flights will open availability of launch pads for Titan III missions.

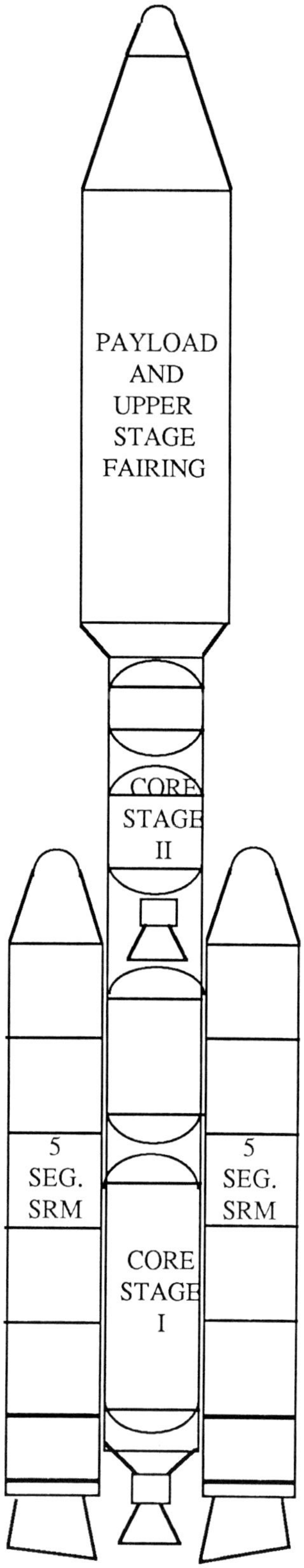

Titan III model as applied to commercial payloads

4-6 Titan IV Improvement Program

Martin Marietta maintains a continuous product improvement program targeted toward specific goals in reliability, performance, safety, schedule dependability, cost effectiveness, and environmental acceptability.

Major near-term hardware goals include flight with a Centaur upper stage, incorporation of the solid rocket motor upgrade; adding a high expansion nozzle to the core first stage engines; spin-formed domes for the propellant tanks; and a composite skirt for the IUS.

Major ground system improvements underway or completed include programmable aerospace ground control equipment; a new facility for solid rocket motor assembly and readiness testing; an upgrade to Florida Launch Complex 40 to accommodate the Centaur upper stage; and a new facility for Centaur stage processing, checkout and storage.

Launch vehicle improvements planned for deliveries beyond the 41 vehicles the Air Force currently has on order will include fault tolerant avionics for the Titan core stages. An upgraded Titan IV will include propellant tanks and structure made of welded lithium-aluminum alloy, which is lighter than aluminum. Composite structures will replace aluminum interstage adapters to save weight. Nested domes will be fitted between the core first stage fuel and oxidizer tanks, eliminating need for a skirt section to join the two tanks.

Future Centaur improvements will center on a redesign that replaces the current twin engines with a single engine that has been upgraded in thrust level. This engine will be fitted with electromechanical actuators for steering, replacing the existing hydraulic actuators. The Centaur avionics system is also undergoing a continuous improvement program directed toward attaining a modern, fault-tolerant system.

A concept for large performance improvements that has been under study for some years is replacement of the Titan core vehicle with a large diameter core. This "Titan V" would use the same propellants as Titan IV and would have a lift capability between 50K and 90K pounds.

The issue of environmental friendliness is a difficult one for the Titan launch vehicles due to the core vehicle being based on highly toxic propellants, and to use of solid propellant rocket boosters that expell a large tonnage of hydrochloric acid in their exhausts. The new solid rocket motor upgrade changes nothing in this area. If Titan V were to be developed, consideration would have to be given to replacing the solid rocket motors with clean-burning solids, or with liquid propellant boosters. But there are no firm plans as yet for a Titan V development.

4-7 Outlook

Titan IV is in the enviable position of being the only expendable launch vehicle in the U.S. inventory capable of lifting heavy payloads into orbit. As the Cold War drew to an end, the Air Force had contracted with Martin Marietta to build 41 vehicles, at a total cost of $10.4 billion. At the time, the Air Force anticipated a flight requirement for approximately ten flights per year.

As viewed by Congress, Titan IV is a new system, recently activated, with scarcely a half dozen flights behind it. Some versions have not flown yet. A large investment has been made to produce an upgraded launch vehicle as well as a modernized infrastructure for launch processing. Proposals to replace the system do not sit well with budget makers and planners inside Washington. So it would seem that an extended future for Titan IV is assured.

But with the end of the Cold War the Air Force's mission requirements dropped off significantly. A part of the diminished demand was due to funding cutbacks that threatened major programs like Milstar. Flight projections dropped from ten per year to seven, and finally to three per year. At this rate, Titan IV's on order will last well into the first decade of the next century. A reduction in flight rate has the effect of increasing launch costs since running costs are spread across fewer vehicles.

High cost is one of Titan IV's drawbacks, and is a leading argument for replacing it with a new system. True costs are difficult to come by. Martin Marietta claims that the launch of a Titan IV approximates $150M, or $200 M with a Centaur stage. The simple arithmetic of dividing 41 vehicles into $10.4 billion places the cost at $250 million. Some estimates out of congressional budgetary sources run as high as $500 million.

A second drawback is the visible difficulty, for whatever reasons, of getting a Titan IV assembled and ready for launch. There have been two occasions when a Titan IV has sat on the launch pad for over a year. Such a delay mandates replacement of the solid rocket motors before the vehicle can fly.

Modernization of the existing fleet is one of the options under consideration by the government, as an alternative to development of a new system. Were this option to be pursued, a significant part of the effort would be directed toward upgrading the Titan system. A decision regarding this approach is likely to be forthcoming in 1994, and will include Titan, Atlas, and Delta programs.

So long as mission requirements remain at two or three launches per year, the Congress is likely to prefer the high attendant costs to the cost of developing a new system.

Lowered Air Force launch requirements provide an opportunity for Martin Marietta to seek other missions for Titan IV. The launch sites will be severely underutilized, having a capacity to launch another seven vehicles per year. One mission for NASA is already scheduled for Titan IV with a Centaur upper stage. Should NASA develop a new personnel carrier to replace the Space Shuttle for transport of personnel to Earth orbit, Titan IV stands as the only U.S. launch vehicle that is capable of launching the craft.

Growth of Titan IV to Titan V is an unlikely development, although it has often been studied. The fundamental objection in this case is doubling or tripling the amount of toxic propellant that would be loaded into the core vehicle. It is unlikely that arguments could overcome environmental concerns in the event of a major mishap on the launch pad.

Titan II appears to be a dead-ended program, although it may go on for twenty years or more. When the inventory of retired ICBM's is expended. the program will cease. Recent purchase of General Dynamics Space Systems Division by Martin Marietta could alter this picture considerably, however, should the company elect to scrap the Atlas boost vehicle and utilize the Titan II's more rapidly.

Titan III is a launch vehicle on hold. Martin Marietta has none in production. The company has the capability to respond to mission requirements that may develop in the Titan III payload range, and plans to do so when opportunities arise in commercial and civil areas.

On August 29, 1994, Lockheed Company and Martin Marietta announced a merger of the two companies, to be known as Lockheed-Martin. The move reflects the continuing retrenchment of the defense industries since the end of the Cold War, and reduced defense spending by the government. Lockheed brings to the new company its small payload launch vehicle family, currently in development. As a result of the acquisition of General Dynamics Space Systems Division and the merger with Lockheed, Lockheed-Martin will be the dominant producer of space launch vehicles in the U.S. for the next decade or two.

5

Space Shuttle and Derivatives

5-1 What is a Space Shuttle?

Some day, perhaps in the not too distant future, it will suddenly dawn upon NASA officials that the agency no longer operates a system called the Space Transportation System (STS). Quite to the contrary, what NASA is operating is a very sophisticated returnable and reconfigurable space station. Not just one, but four!

For all practical purposes , the system ceased to be a STS during the stand-down following the Challenger accident in January of 1986. NASA was directed to terminate the business of transporting commercial payloads into orbit, and the Air Force began the process of converting its payloads for transport by expendable launch vehicles. Transportation of payloads on the Shuttle, including spacecraft, is now almost entirely confined to specialized NASA objectives.

This altered view relieves NASA of the strain of having to compare the Space Shuttle with expendable launch vehicles. As a launch vehicle, the cost per shuttle launch is prohibitively high, if prorated across the annual Space Shuttle budget. But viewed as a returnable Space Station, cost comparisons with expendable launch vehicles become groundless. Judged on its own merits as a returnable space station, Space Shuttle may in the long run turn out to be one of the nation's best space bargains.

Apparently unaware of what has in fact occurred, the Shuttle program is concentrating on system modifications that can extend stay-time in orbit for thirty or more days. This is quite a departure from the original concept of a machine whose main purpose was a brisk business hauling payloads into low earth-orbit. Nothing is happening to retrieve its original function. Everything is happening in the direction of improving the shuttle's capability as a returnable space station.

The International Space Station will be a fixed installation, located permanently in an extremely limited range of orbits. Flexible space stations Atlantis, Endeavor, Columbia, and Discovery can go into many orbits, can be altered in configurations and missions, will have respectable stay-times in orbit, and have the added capability of servicing the Space Station as needed until automatic servicing spacecraft are created.

With five functioning space stations by the end of this century, the United States will be embarrassingly rich in low-earth orbit research capacity. If the naysayers in the public and the Congress can be held at bay, the window may soon be opened to a new and productive era in beneficial space exploitation, with the United States continuing to lead.

5-2 Space Shuttle's Origin and Description

During the 1960's, a series of studies were undertaken by government and industry, aimed at developing fully reusable space launch vehicles. The consensus among spacecraft designers and managers was that space launch costs had to be reduced and the way to do it was with flyback boosters and orbiters that would last a hundred or more trips.

NASA Langley had by this time already experimented with several lifting body shapes for winged return from Earth orbit. The Air Force had made significant progress with their Dyna-Soar vehicle, a 10,000 pound single passenger winged vehicle that was an essential part of a manned military orbital laboratory. The Dyna-Soar program was cancelled in 1963 by Defense Secretary Robert McNamara for lack of a defined role. It would have flown two years later, about twenty-five years before the first Space Shuttle flight.

In July of 1970, NASA issued two Phase B contracts for industry studies of an integrated boost and orbiter flight system. Ambitious ground rules for the study included stipulations that the shuttle should be a fully reusable two stage vehicle. It was to launch vertically but land horizontally on landing fields. Turnaround time was not to exceed two weeks and a flight rate of up to 75 launches per year could be expected. The contractors selected for Phase B included a team of McDonnell Douglas and Martin Marietta, and Rockwell International, with General Dynamics as a subcontractor.

The Phase B studies produced several concepts for fully reusable launch vehicles, all of them ambitious and costly undertakings. Enthusiasm cooled, however, when in May of 1971 NASA was informed by the OMB that prospects of budget increases to accomplish the project in the next few years were zero. If NASA wanted the Space Shuttle, it had to figure ways to accomplish it within the existing budget, and still maintain a balanced science and applications program.

Literally hundreds of concepts were examined in the ensuing months, all of them centering on a recoverable orbiter and expendable boosters. Both liquid-fueled boosters and solid rocket boosters were examined. By December of 1971 NASA was ready to adopt a parallel burn design that had been promoted by the Grumman Company and favored by Johnson Space Flight Center's chief engineer, Max Faget.

The selected concept was a hydrogen-fueled orbiter with its propellants fed from a jettisonable external tank, with boost thrust provided by two solid propellant rocket motors. All engines were to light off on the launch pad. As a step toward retaining some of the reusability aspects of the original concepts, NASA planned to recover the expended solid motor shells and reuse them after refurbishment.

Space Shuttle lifts off at Florida's Kennedy Space Launch Center (NASA)

The Space Shuttle is comprised of three major subsystems: the orbiter; the disposable propellant tank that fuels the orbiter's rocket engines, and twin solid rocket boosters (SRB's). The orbiter was designed and built by Rockwell International's Space Transportation Division of Downey, California. The propellant tanks are produced by Martin Marietta at NASA's Michoud Assembly Facility in Louisiana. The solid propellant rocket motors, described in Chapter 7, are built by the Wasatch Division of Thiokol Chemical Corporation, Brigham City, Utah.

The orbiter is a double-delta winged vehicle capable of carrying passengers and cargo to low-earth orbit and back to a controlled, unpowered glide landing.

The Shuttle fleet is based on four operational orbiters. Four flight articles were originally built, but one was lost in the Challenger accident. A replacement, Shuttle Atlantis, was built and recently placed into service. External tanks and boosters are produced in quantities to satisfy flight schedule demand.

The orbiter ranks among the most complex spacecraft ever built. It is required to operate under extreme thermostructural load during its duty cycle that begins with launch, acceleration at high g's to orbital velocity; thermal cycling from the heat of the sun to the cold of the Earth's shadow many times during a mission; earth atmosphere reentry through extreme temperature buildup due to aero heating; and landing at an airport like an aircraft without thc benefit of jet power.

For main propulsion power, the orbiter has three engines built by Rocketdyne Division of Rockwell International. The Space Shuttle Main Engines (SSME's) are ignited before liftoff to verify satisfactory operation, following which the boosters are started. The engines operate at very high pressure, which allowed designers to achieve reasonable performance at sea level while maximizing performance at altitude.

A second propulsion device that is completely contained within the orbiter is the Orbit Maneuvering System (OMS). The OMS provides thrust for orbital insertion after the main engines have been shut down and the external tank has been jettisoned. It also provides thrust for orbit circularization, orbit transfer, rendezvous and de-orbit. The OMS is contained in two aft pods on either side of the vertical stabilizer.

Within the OMS pods and distributed elsewhere on the orbiter are reaction control rockets for control and stabilization when in orbit. There are primary thrusters which operate at 870 pounds thrust and secondary, or vernier thrusters, which operate at 25 pounds thrust.

Propellants for the OMS and reaction control systems are nitrogen tetroxide and mono-methyl hydrazine.

Space Shuttle main engine produced by Rocketdyne Division of Rockwell International

Other major elements of the orbiter are a fuel cell power supply , an avionics system that is two-failure tolerant, a crew compartment, a life support system capable of sustaining five or six crewmen for more than a week, thermal radiators for dissipating heat generated within the orbiter, and payload handling hardware, including a remotely operated arm for holding and maneuvering space hardware.

The orbiter main structure is mainly of aluminum alloys. Major structural elements are the crew compartment, mid-fuselage, aft-fuselage, OMS pods, wings, and vertical stabilizer. Fabrication of these elements was distributed among a number of subcontractors. Assembly of the orbiters was done in Rockwell's Palmdale, California plant.

The body is insulated from aero-heating by a combination of reusable tiles which are glued on, insulation blankets, and localized use of advanced composites.

Orbiter launches take place from complexes at Kennedy Space Center in Florida. A launch site at Vandenberg Air Force Base in California was nearing completion, but work was terminated when a decision was made to pursue a more modest program with the Shuttle, following the Challenger accident. Shuttle landings take place either at Edwards Air Force Base in California or at the Shuttle landing field in Florida, depending on weather conditions.

The Space Shuttle external tank (ET) contains liquid oxygen and liquid hydrogen for fueling the orbiter's three main engines. The ET measures 153.8 feet in length and 27.6 feet in diameter. It consists of a forward oxygen tank and an aft hydrogen tank, joined by an intertank cylindrical adapter. The oxygen tank contains a baffle structure to inhibit sloshing of propellant during flight. Baffles are not required in the hydrogen tank since sloshing of this light weight propellant does not induce significant disturbing forces. Propellant and electrical interconnects with the orbiter are located near the aft shuttle attachment structure.

The entire external tank assembly is built of fusion welded aluminum alloy. The ET also carries forward and aft structural attachment points for the orbiter and two solid rocket boosters. The tanks are covered with a foam insulation to reduce propellant boiloff and to prevent condensation of liquid air on the super cold surface of the hydrogen tank.

Since the external tank nearly reaches orbit, its empty weight corresponds closely to payload weight. It was determined early in the program that efforts to reduce the ET weight were needed. The program now flies what is termed the "lightweight" ET. This tank has an empty weight of 66,800 pounds, about 10,300 pounds lighter than the original tank. A new model, saving an additional 8,000 pounds, made of aluminum/lithium alloy, is currently under development by NASA to enable launching the heavy components of the international space station beginning in 1997.

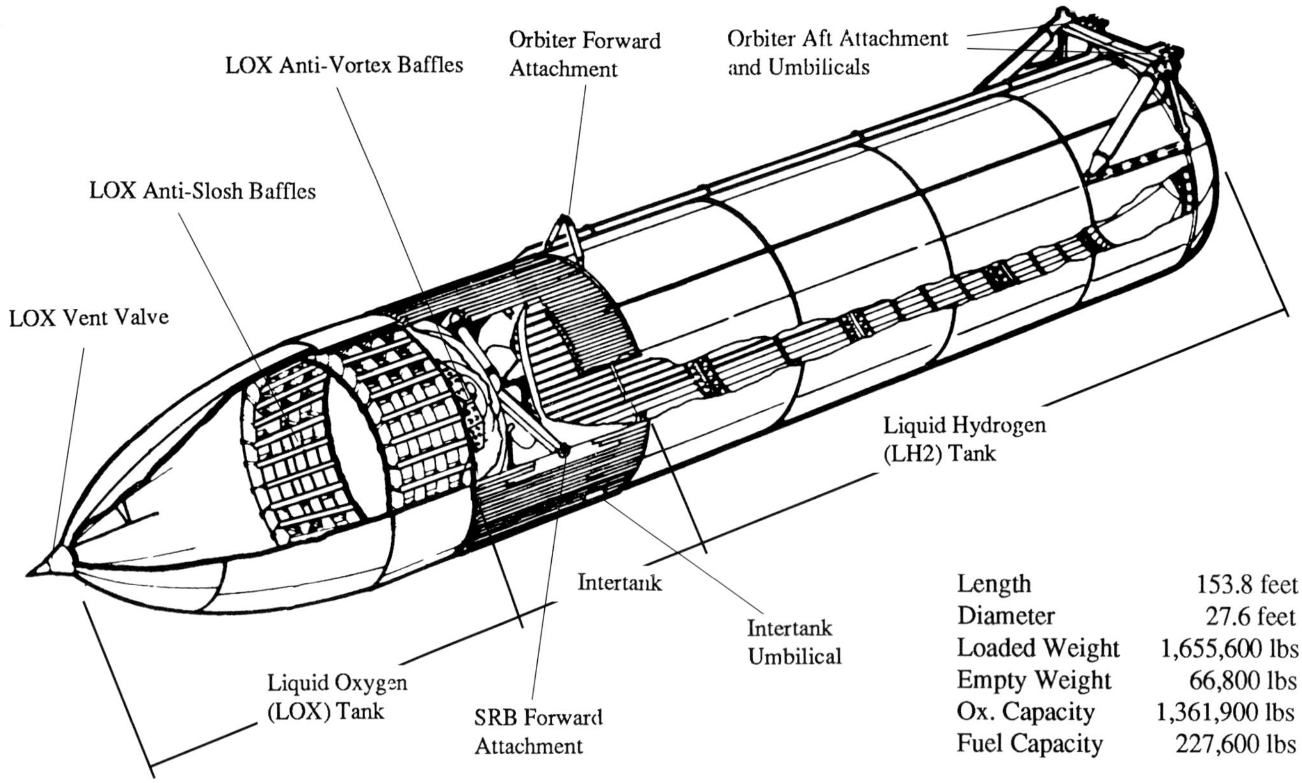

Length	153.8 feet
Diameter	27.6 feet
Loaded Weight	1,655,600 lbs
Empty Weight	66,800 lbs
Ox. Capacity	1,361,900 lbs
Fuel Capacity	227,600 lbs

The Space Shuttle light weight External Tank (NASA)

5-3 Shuttle Operations

After fifty plus flights, the shuttle has experienced only one failure, yielding a reliability above 98%. A part of this success can be attributed to the consistent routine followed for every flight. A nominal flight begins with the solid rocket boosters being stacked, segment by segment, on the Mobile Launch Platform in the Vertical Assembly Building (VAB). Meanwhile, the orbiter is being serviced in the Orbiter Processing Facility.

The external tank is erected next and mated to the stacked SRBs. The orbiter is then moved to the VAB and connected to the external tank. The orbiter may already have the payload installed in its cargo bay, or installation may be delayed until the system is on the launch pad. After checkout and installation of range safety ordnance, a crawler transporter moves the Space Shuttle system to the launch pad, about 3.5 miles distant. There, servicing connections are made and the Shuttle is prepared for launch.

If the payload was not installed in the Orbiter Processing Facility, it would be installed at the launch pad following prelaunch activities. The orbiter is serviced with propellants and, in the final hours, the external tank is filled with liquid oxygen and liquid hydrogen. Final checks are made and the system is cleared for main-engine ignition.

The Shuttle main engines are ignited first and verified to be operating at design thrust levels. This takes place in a matter of seconds. A signal is then sent to ignite the SRB's, followed immediately by firing the eight explosive hold-down bolts that secure the SRB's to the launch platform.

Approximately 120 seconds into flight, the SRB's burn out and are separated from the ET. This is done by firing explosive bolts to sever the connections, augmented by firing eight small solid rocket motors to ensure a clean separation from the vehicle. The expended SRBs follow a trajectory that lands them in the ocean about 140 miles downrange, where they are recovered by a ship and towed back to Cape Canaveral for refurbishment.

The orbiter and external tank continue their ascent, powered by the three Shuttle main engines. About eight minutes later and just short of attaining orbital velocity, the main engines are shut down and the ET is jettisoned. The ET continues on a ballistic trajectory, reentering the atmosphere over the Indian Ocean, where it disintegrates.

The OMS engines on the orbiter are then ignited to accelerate the orbiter into an elliptical low earth-orbit. A second burn at apogee circularizes the orbit. On direct insertion missions, only one burn is required. Orbital altitudes vary between 140 miles and 380 miles, depending on mission requirements. Shuttle payload weight-carrying capability drops off significantly with increasing altitude.

When the mission is completed, the Shuttle's reaction control system is used to turn the vehicle tail forward. The two OMS engines are then fired to slow the orbiter for de-orbit. The RCS is then used to turn the Orbiter nose forward for reentry, which occurs at an altitude of about 400,000 feet, 5000 miles distant from the launch site. The Orbiter's angle of attack is forty degrees and its velocity is about 17,000 miles per hour at reentry. RCS thrusters are used to control the Orbiter until dynamic pressures are sufficiently high for control surfaces on the Orbiter to take over.

During reentry, an enormous amount of energy must be dissipated, which is manifested in high heating on the orbiter's surfaces. The brunt of the load is taken by the insulation tiles which cover the areas most affected by aero-heating.

The approach and landing phase begins at an altitude of about 10,000 feet and a velocity of 320 miles per hour, roughly eight miles from touchdown. Auto-land guidance is initiated at this point. The orbiter descends at a minus 19 degree glide slope, aimed at a target one mile in front of the runway. At about 1700 feet altitude a pre-flare maneuver is initiated to position the craft for a 1.5 degree glide slope in preparation for landing. In the last moments the orbiter's sink rate is reduced to less than 9 feet per second and touchdown occurs about 2500 feet past the runway threshold. Landing speed is 220 miles per hour.

If the orbiter is landed at KSC, it is moved directly to the Orbiter Processing Facility. If the landing is made at Edwards Air Force Base in California, the orbiter is safed on the runway and is then moved to a facility where it is inspected, then loaded on the transport aircraft for ferrying back to KSC.

With the cancellation of the Shuttle launch site at Vandenberg Air Force Base, the U.S. remains without a manned polar launch capability. Polar launches from KSC cannot be made due to launch path restrictions that prohibit overflight of inhabited land masses during flight ascent.

5-4 Space Shuttle Improvements

Improvements in the Space Shuttle are concerned with performance upgrades, better safety and reliability, more efficient turnaround operations, elimination of component obsolescence, and lower cost. Rockwell International has a continuing program of government funded and internally funded studies that look to continued use of the system in the nation's space program for 25 years or more. Improvements in systems provided by other companies are also a part of the continuing improvement program.

In the aftermath of the Challenger accident, NASA was forced to make design fixes to correct safety deficiencies in the solid rocket boosters and in the orbiter. The changes caused a significant a payload penalty which can be retrieved, along with considerable added gain, when downstream improvements involving the ET and SRBs are incorporated.

Martin Marietta has already produced what is termed the "lightweight" external tank, which has been in service for several years. The lightweight tank weighs 66,000 pounds. But another ten to fifteen percent reduction in weight can be obtained if tank manufacturing is converted to use of an aluminum-lithium alloy, also developed by Martin Marietta. Manufacturing forming and welding process are now sufficiently advanced that production of new alloy tanks can begin in the near future. The new alloy is far more expensive than the aluminum presently in use, but when one considers that each pound saved is about equal to a pound of payload worth up to $10,000, any increase in tank materials cost is trivial.

A second area of planned improvement concerns the Shuttle's booster rockets. Repairs to the existing solid rocket boosters following the Challenger accident were viewed as a temporary expedient. NASA was directed by Congress to pursue design and development of a more reliable replacement. A new booster would not only be more reliable, but would regain payload lift capability which had been eroded by safety modifications related to the Challenger accident. A major effort development effort was undertaken by NASA, designated the "Advanced Solid Rocket Motor" (ASRM) program. More about the ASRM in Chapter 7.

NASA has also undertaken several studies directed toward replacing solid rocket boosters with liquid-fueled rockets. Proponents of liquid-fueled boosters cite the advantage of having safe abort modes since, unlike SRBs, liquid-fueled rockets can be shut off if required. Other advantages are safer ground operations, increased payload, reduced launch costs, improved mission flexibility, and environmentally benign exhaust products. Most concepts for liquid-fueled boosters use liquid oxygen and RP-1 propellants. Candidate rocket engines for such boosters include the Apollo F-1 booster engine, Rocketdyne's proposed STBE engine, fueled by liquid oxygen and liquid methane, and the Russian Energia rocket engine, promoted by Pratt & Whitney Company under an exclusive import agreement with the Russian manufacturer.

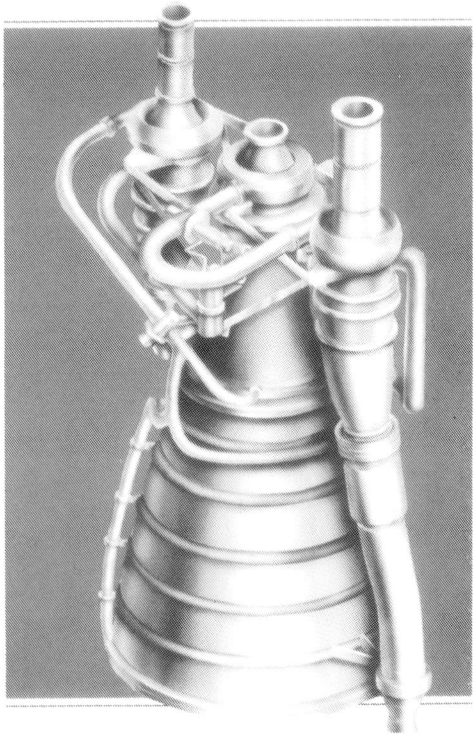

Rocketdyne's proposed space transportation boost engine

Rocketdyne's F-1 boost engine, used on Saturn V

Russian RD-170 four-barrel oxygen/kerosene engine

The current Space Shuttle baseline performance is set at 58,000 pounds payload delivered to 150 nautical miles altitude, 28.5 degrees inclination. Predicted performance with ASRMs replacing the SRBs is pegged at 70,000 pounds. Substitution of a new lightweight external tank would raise the payload to 80,000 pounds. Adoption of liquid rocket boosters in place of the ASRMs would gain another 5,000 pounds.

Besides a continuing program to improve the operation of Shuttle orbiters through component upgrades, Rockwell engineers are studying other growth options, notably directed toward extending mission stay-time. Extended duration orbiters basically require more of the same provisions that initially allowed them a seven day stay-time. These include additional cryogenics for fuel cell power generation and oxygen supply, pressurized nitrogen tanks, a regenerative CO2 removal system to replace existing lithium hydroxide CO2 extraction devices, and additional orbiter stowage capacity.

Space shuttle Columbia contains the above modifications, enabling it to stay in orbit for 16 days, with a two-day contingency. The remaining three Shuttles could be modified similarly when returned to Rockwell's Palmdale plant for major overhaul, although all are already equipped to some degree to accommodate selected extended-duration installations.

Rockwell's plans for extended-duration missions do not stop with the 16 day modifications. A 30 day mission capability appears to be feasible with nine or ten relatively minor changes. The main benefits of extended duration are the opportunities to accomplish more life science and materials processing during any given mission.

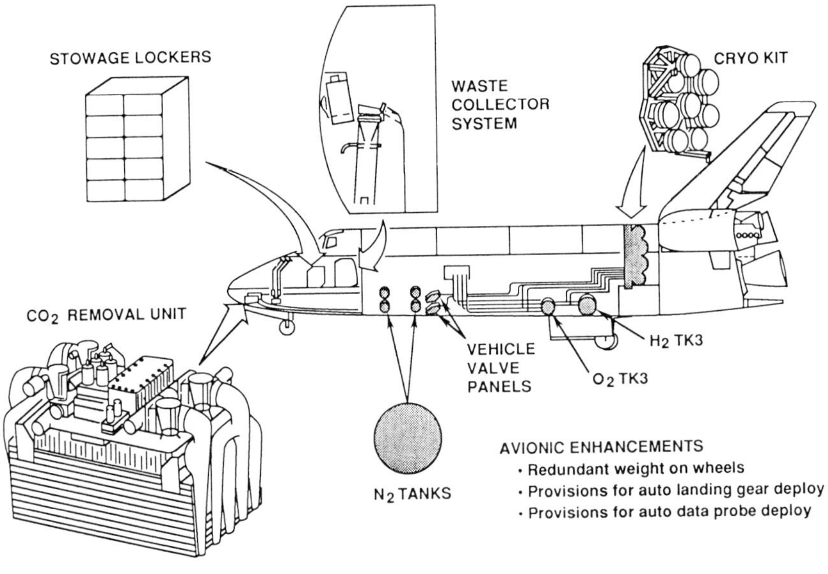

Kit Modifications for the Extended Duration Orbiter (Rockwell)

Rockwell is also is working on plans for a crewless orbiter that would be used to carry cargo to low-earth orbit. The plan is to use existing vehicles equipped with an automated orbiter kit that is installed at mid-deck during preflight processing. An automated orbiter based on this approach is predicted to have a 12,500 pound payload performance gain.

A second approach is a new vehicle build, which would be configured first as a cargo carrier , but would be retrofittable to fly with a crew. This new orbiter would have a 20,000 pound performance gain.

It is a matter of speculation whether such devices could compete with expendable launch vehicles as payload carriers. Success of the concept would probably hinge on NASA's own future requirements for carrying heavy payloads into orbit. Current policy forbids transportation of commercial payloads and the Air Force has turned to using its own resources for flying military payloads.

In one scenario, however, an automated orbiter may be a good idea. In the event that some venture in space manufacturing or space processing proves to be highly successful, one can imagine the use of automated orbiters to haul little factories into orbit where they do their work, then return them to Earth for product recovery and reload for another cycle.

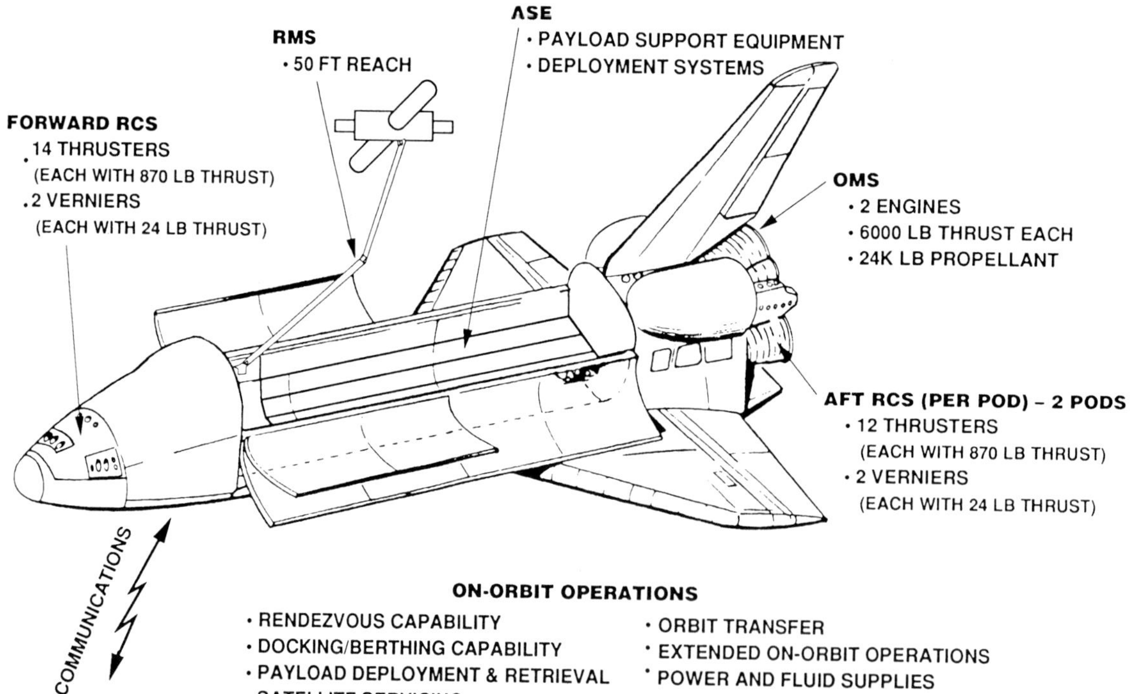

Rockwell's concept for an automated orbiter provides the same on-orbit capability as an orbiter with a crew. (Rockwell)

5-5 Shuttle II

Since budget considerations resulted in an admitted compromise in the Space Shuttle design, it was only natural that efforts would continue along lines leading to definition of a next model vehicle, or Shuttle II. Studies within NASA and among contractors to define the next generation Shuttle were begun even while the Space Shuttle was still being designed.

A joint study performed by NASA Langley and the George Washington Institute for the Advancement of Sciences, completed in July 1989, produced a number of vehicle types for consideration.

Shuttle II, as NASA engineers conceived it, was to make maximum use of technology advances that had escaped the original Space Shuttle designers, for whatever reason. The perceived structures made use of carbon composites and featured more durable, reusable external insulation surfaces. Containerized cargo systems and adoption of modern, line-replaceable aircraft components were featured in the evolving concepts. Shortcomings of the existing Space Shuttle, as identified by the using astronauts, were also considered.

Imitating some of the earlier concepts for a totally reusable launch vehicle, the final Shuttle II in Langley studies settled on a three-part system consisting of a large reusable orbiter that carried its own propellants, and a pair of flyback, reusable boosters. Three different sizes of vehicles were presented at the conclusion of the studies, with payload capabilities ranging from 37,000 pounds to 84,000 pounds.

With the exception of the largest version of the Shuttle II presented in this study, the boosters used the Space Shuttle Main Engine, which has a sea-level rated thrust of 375,000 pounds. The largest boosters were fueled with liquid oxygen and the hydrocarbon fuel RP-1. Propulsion was provided by engines that were at that time under study by Aerojet and Pratt & Whitney. The proposed engine had a projected sea-level thrust of 625,000 pounds and three were used on each booster.

A unique feature of the boosters was an oblique arrangement that facilitated close integration of the three vehicles on the launch pad. The wings were to be rotated into position after burnout and separation, allowing an unpowered glide return to the launch site. Oblique wings were flight tested by the Dryden Flight Research Facility during the 1980's with a view to providing commercial aircraft configurations that are more fuel efficient.

The unmanned boosters were aerodynamically similar to the orbiter, but smaller. The propellant tanks and simple skin-stringer structure were all planned to be built of aluminum alloy. The aerodynamic heating loads proved to be low, since the boosters staged at about Mach 3. The tanks and structure were to function as a heat sink against aero-heating.

The proposed orbiters are very large delta wing vehicles, a consequence of their high propellant capacities. The main body has a circular cross section with an ogive shaped forebody. The crew compartment, located in the mid-body between the oxygen and hydrogen tanks, has no raised canopy for pilot visibility. During approach and landing, a remote television camera is deployed to give the pilot visibility. Flush mounted view ports around the crew compartment give the crew exterior viewing capability.

The orbiters feature a sharply truncated delta wing to give the vehicle lift during reentry and landing. The wing tips are equipped with vertical stabilizers as a secondary method of directional control. Primary control is achieved with a dorsal stabilizer mounted on the forebody. A combination of the forward dorsal stabilizer and small aft-vertical stabilizers at the wing tips was found to minimize drag during ascent as well as structural weight.

The orbiters in all three concepts rely on three Space Shuttle Main Engines, which have a sea-level thrust of 375,000 pounds each. Future re-examination of the concept would in all likelihood recommend adoption of the new space transportation main engine (STME) that NASA has under investigation for possible joint use with the Air Force in the next generation expendable launch vehicle.

A number of other studies have also investigated possible configurations for a Shuttle II. Some are entirely reusable and others still use expendable components. They are merely exercises in preliminary design, hoping for a discovery that would shake the foundations of the space transportation industry.

Fully reusable Space Shuttle studies continue at a low level within NASA and in industry, although it is universally recognized that the likelihood of a start-up for a Shuttle II is practically zero. NASA plans for operating the current Shuttle extend to the year 2020, and a replacement system may completely skirt Shuttle II concepts in favor of a much smaller crew carrier, launched by an expendable rocket. In such an event, payloads would continue to be launched with expendable vehicles. The current Space Shuttle may be the first and last of its genre, a victim of unexpectedly high cost of operation.

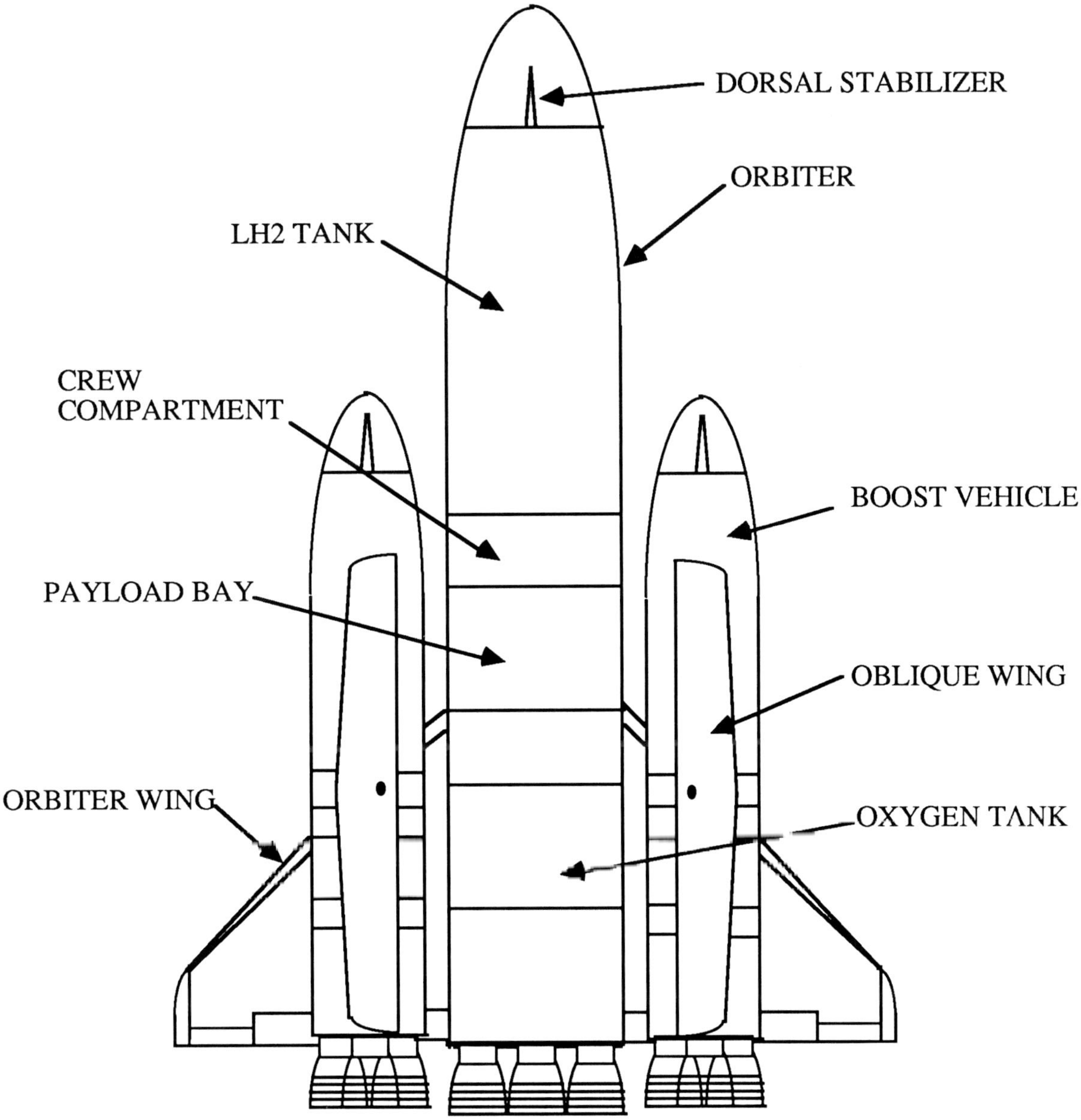

PAYLOAD	ORBITER HT.	BOOSTER HT.	ORBITER ENGINES	BOOST ENGINES
37,000 LB	148 FT.	116 FT	3 SSME	4 SSME
70,000 LB	169 FT	133 FT	3 SSME	4 SSME
84,000 LB	197 FT	126 FT	3 SSME	3 LO2/RP1 @ 675 K LB

The Proposed Langley Research Center Shuttle II

5-6 Shuttle-C

As early as 1976 studies were underway in industry to explore methods of using shuttle derived hardware for launching heavy payloads into orbit. If a cargo container were to be substituted for the orbiter, which weighs approximately 200,000 pounds, then in an unmanned configuration Shuttle hardware could be used to launch payloads weighing up to 150,000 pounds. The idea was attractive because the heavy payload carriers could be launched from the same pads as the Space Shuttle.

A large number of configurations were investigated over several years with the bulk of the effort undertaken initially by Boeing and finally by Martin Marietta. During the early studies, one basic configuration prevailed. It featured a side-mounted cargo container positioned on the Space Shuttle mounts, with a propulsion pod similar to the Shuttle's. Due to the high cost of Space Shuttle main engines, attempts were made to configure recoverable propulsion/avionics pods (P/A module) that could be used for many flights. NASA eventually awarded separate contracts to investigate feasibility of recovering the propulsion pod from orbit.

The unmanned Shuttle entered into joint NASA/ Air Force studies in the late 1980's, when NASA proposed the Shuttle-C as a candidate for the Heavy Lift Launch Vehicle (HLLV) the Air Force was seeking. Shuttle-C was ruled out by the Air Force because it could not offer the dramatic reduction in cost to access space that the Department of Defense was demanding. This left NASA free to promote the Shuttle-C as a vehicle for meeting its own heavy payload requirements.

NASA's driving requirement for a heavy lifter centered on plans to launch the larger elements of Space Station Freedom into orbit. In the background, however, were visions of other adaptations that could serve the proposed Space Exploration Initiative Lunar-Mars missions. Whereas the high energy Centaur upper stage was abandoned as being unsafe for transport into orbit by manned vehicles, it could once again be considered as part of an unmanned vehicle, serving lunar and planetary missions.

In November of 1987 Marshall Space Flight Center awarded contracts to Martin Marietta, United States Booster Co., and Rockwell International to define a Shuttle-derived launch vehicle. More configurations were created and compared. One configuration placed a propulsion pod at the base of the external tank, with a large payload compartment at the forward end. This arrangement fell out of consideration due to extensive launch pad and assembly building modifications that would be required due to its great height. The launch pad would also have required modification to accommodate the altered engine exhaust path.

Based on the results of these studies , Marshall Space Flight Center released a request for a proposal to industry for development of a Shuttle derived launch vehicle with a side-mounted cargo element. The baseline payload was specified as 15 feet in diameter by 72 feet in length, weighing 103,000 pounds. The reference mission was a 253 mile orbit at 28.5 degrees. Use of existing SRB's and the external tank were specified. The cargo element was to be fitted with a modified Shuttle aft-fuselage with the vertical stabilizer, body flap, and one Shuttle main engine removed. Avionics and engines were required to be contained within the aft fuselage in anticipation of a possibility that this unit, called the Propulsion/ Avionics Module, could be recovered.

A contract was awarded to an industrial team consisting of Martin Marietta, Boeing, Teledyne, Brown Engineering, Intermetrics, and United States Booster Inc., in February of 1988. During the ensuing eighteen months, design studies resulted in numerous changes and refinements. Initial plans specifying recovery of the P/A module were abandoned, although it was retained for possible use on a second generation vehicle, termed Shuttle-C Block II. This vehicle featured a larger cargo bay and used three Shuttle main engines. It could carry up to 170,000 pounds of payload to low-earth orbit.

A still larger version of Shuttle-C, Called Shuttle-Z, a 300,000 pound payload behemoth, also appeared as one of the future Shuttle-derived alternatives.

Shuttle-C's future ran aground in the Congress as a result of steeply escalating costs of Space Station Freedom. The program was cancelled in 1990.

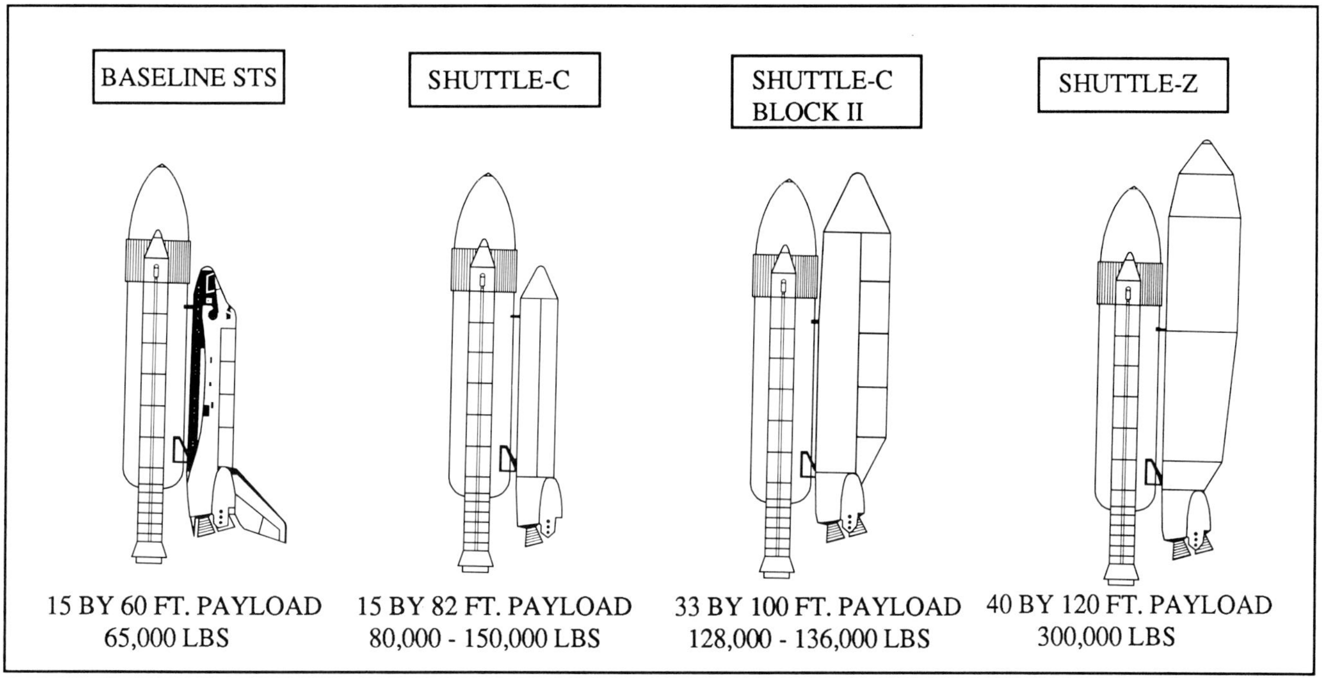

Shuttle Derived launch vehicle alternatives

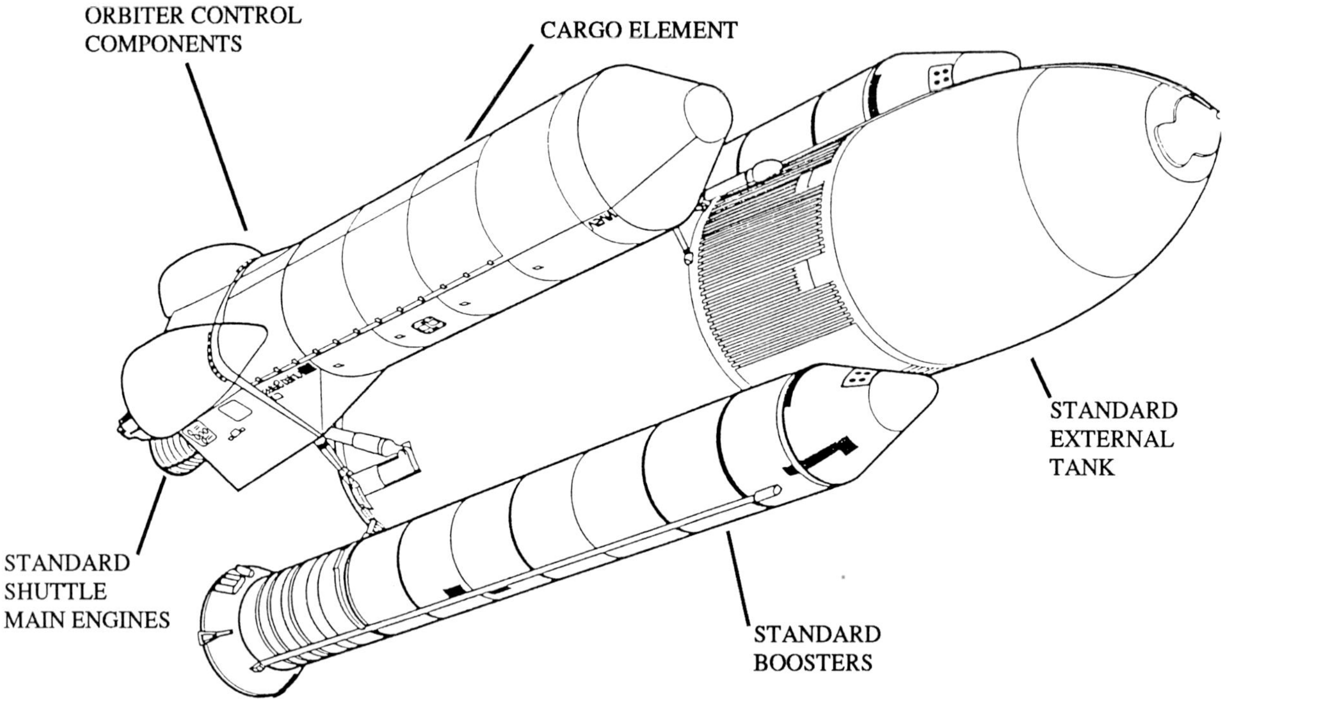

Shuttle-C, as it appeared when the program was cancelled in 1990 (United Technologies)

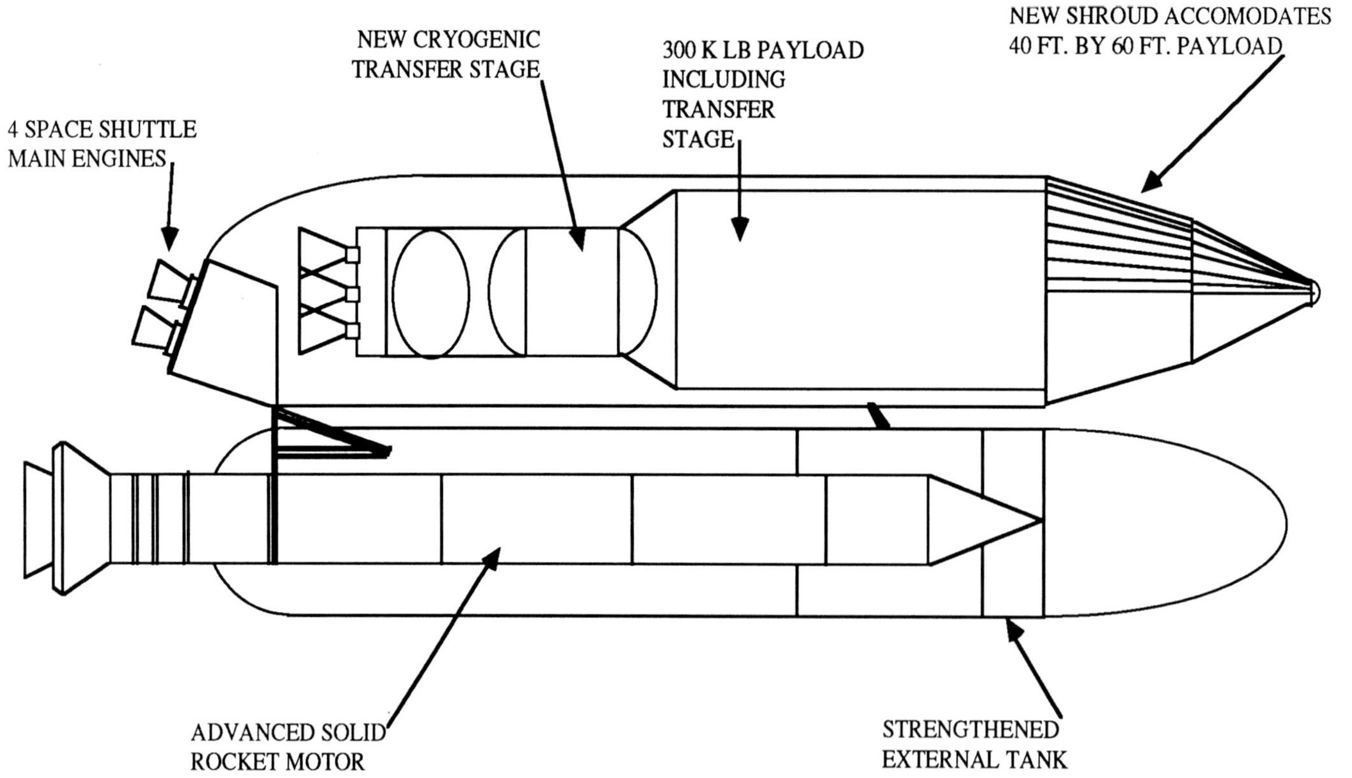

Shuttle-Z, a heavy lift launch vehicle defined during Shuttle-C studies

Among the advantages quoted for Shuttle-C was the high commonality with components of the existing Space Transportation System, leading to the conclusion that the vehicle could be economically and rapidly developed. An interesting turn of events was a decision to make the Shuttle-C capable of launching the Centaur high-energy upper stage, previously deemed unsafe to carry in the Space Shuttle cargo bay. But instead of the wide-body Centaur originally planned for the Shuttle, the same configuration of Centaur being designed for Titan IVwas selected for Shuttle-C.

Estimates of the cost of development of Shuttle-C ranged from $985 million to $1.8 billion. The cost to launch was estimated at $424 million, resulting in a cost-per-pound of payload into orbit of $4,240. For comparison, cost to orbit a pound of payload on Titan IV is about $5,000. A method proposed to keep the cost of Shuttle-C's as low as possible was to use Space Shuttle main engines that are approaching the end of their service life. Thus, it could be argued that the cost of engines for Shuttle-C would be essentially nil.

NASA's planners foresaw a need for flying 14 orbiters per year and 10 Shuttle-C's beginning in 1995. At this rate the existing supply of used engines would be rapidly depleted, forcing NASA to purchase new engines. At a cost of $38 million per engine, the launch cost of a Shuttle-C would escalate to well over $500 million, eliminating the claimed cost advantage over Titan IV. To all appearances, Shuttle-C would cost about the same to fly as the Space Shuttle, although carrying a much heavier payload.

Shuttle-C progressed as far as construction of an engineering development model by the Boeing Company. This model was intended to be used for fit-and-function checks for the various ground processing operations planned for Shuttle C.

The 1990 decision to cancel Shuttle-C effectively concluded all work on the concept. It remains, however, a leading candidate for lifting heavy payloads, should there be only an occasional requirement in future space ventures.

5-7 The HL-20 Personnel Launch System

NASA has returned to the lifting-body shape in its search for a personnel transportation system that would be more economical than the Space Shuttle. Future ferrying of astronauts to Space Station Freedom with the Space Shuttle is a problem of excessively high cost and would present a considerable strain on the Space Station's operating budget.

Engineers at the Langley Research Center (LaRC) have created the HL-20 Personnel Launch System (PLS) as part of a NASA effort to satisfy Space Station requirements. Other NASA centers, notably Ames and Johnson Space Center are also studying the problem.

The HL-20's overall design is constrained by a NASA requirement that the system be capable of fitting into the cargo bay of a Space Shuttle Orbiter. A part of its mission was to serve as a crew return vehicle at the Space Station, in the event of an emergency. To do this, the HL-20's wings can be folded, resulting in an overall width of 13.53 feet. Overall length is 28.24 feet and estimated gross weight is 24,000 pounds.

The current design accommodates a two-man crew and up to eight passengers. Passengers occupy the cargo space in four rows of two. When no passengers are carried, the PLS can carry up to 4000 pounds of payload.

The HL-20 design was begun in mid-1989. For the next two years, extensive wind tunnel testing was performed as part of an effort to predict the aerodynamic characteristics of the craft over a speed range of Mach 0.05 to 20.3 The results led to a design with an L/D of 4.2:1, which provided good lift and permitted the lightweight vehicle to reach up to a 1000 mile cross-range capability. The wind tunnel tests showed good stability throughout the flight regime.

Primary candidates for launching the Personnel Carrier were at first considered to be Titan IV and a National Launch System (NLS) vehicle. The NLS is presently defunct, having suffered funding cutoff by the Congress.

This leaves the Titan IV as the sole candidate. Titan IV is not man-rated, although a version of the core vehicle was man-rated for the Gemini program in the 1960's. But Titan IV remains a high cost system burdened by long flight preparation time.

An appealing way of launching the PLS is on a new hybrid launch vehicle. Hybrids are safe rockets, ideally matched to personnel transport. Moreover, cost predictions show availability at a quarter the cost of Titan IV.

Interior view of PLS seating up to ten passengers (NASA)

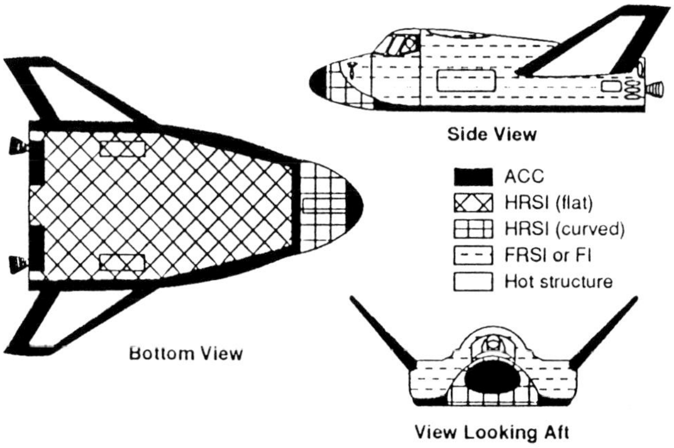

HL-20 shape and exterior

The Langley Research Center HL-20 Personnel Launch System (NASA)

5-8 Outlook

High cost of operations has produced many Space Shuttle program detractors in the general public and in the Congress. As a launch vehicle, despite its high reliability, the Shuttle is a dismal failure. At a launch rate of six to eight flights per year, estimates of per flight costs range from $600 million to $1billion. Such estimates are drawn from knowledge that the overall Shuttle program costs about $6 billion per year.

But as a returnable, reconfigurable space station, the Space Shuttle is a significant success. A wiser Congress would probably realize that conversion of the Shuttle fleet to extended stay-times would probably permit achievement of most of the objectives of ,the International Space Station, at lower cost and with considerably more flexibility. A wiser NASA, on the other hand, should realize that the agency is flying space stations, not launch vehicles, and adjust its stance accordingly.

The Shuttle will continue to be used for a decade or two. There is nothing to replace it, and in the face of budget deficits the Congress will be slow to authorize new system starts for NASA. NASA's budget is not likely to grow soon, and it is already heavily burdened with Space Station Freedom.

Shuttle-C remains a viable idea, probably the most economical way for the country to achieve a moderately heavy-lift capability. Its future hinges on appearance of a requirement for a few heavy payloads sometime in the future. If the requirement is for many payloads, the nation would be better off developing a new heavy-lift launch system, striving for low cost in the process. The heavier payload versions of Shuttle-C are also attractive in a situation where a few payloads are required to be launched of a weight greater than can be managed by a Shuttle-C.

The Personnel Launch System has been viewed as a rescue device for Space Station Freedom, in the event that astronauts are required to evacuate the base. In this situation, its future is threatened by a planned adoption of the Soyuz module by the Space Station program.

In the larger view, the transportation of personnel to and from the Space Station must be viewed as a separate requirement which only the Space Shuttle can now fulfil, but very expensively. If the manned space program is going to move forward in the coming years, the nation will need a vehicle like the PLS to provide the resilience that is mandated in the event of a shuttle being disabled in orbit, or a need arises to respond rapidly to a developing situation at the Space Station.

The PLS is slowly building a base for its support. It will probably be a funded program before the turn of the century.

6

Small Payload Launch Vehicles

6-1 How the Requirement Developed

During the 1980's, scientists at the Defense Advanced Research Projects Agency (DARPA) undertook to apply advancements in microelectronics to construction of miniaturized spacecraft, containing the same functions as their larger cousins. Impetus for this effort stemmed from projections of lower launch costs as well as lower costs for the spacecraft themselves. Development efforts, undertaken jointly with industry, were successful, marking the beginning of a promising new industry.

While the original direction of the research effort was directed toward benefiting military applications, an immediate interest grew in applying small spacecraft to commercial communications. Before the end of the decade, models of small spacecraft were appearing at aerospace industry conventions across the country.

Concurrently, a requirement was developed for small launch vehicles to launch the spacecraft. DoD inquiries were then directed toward procuring launch vehicles with a payload capability ranging from 900 to 1500 pounds to low-earth orbit.

The principal commercial application of interest for small spacecraft is in mobile satellite communications, a proposed system that enables point-to-point telephony anywhere on earth. Such systems employ satellites in low-earth orbit, as opposed to the conventional systems which are located in geostationary orbits .With satellites in low-earth orbit, communicating ground stations can be small and portable.

Because a low-earth orbit satellite is visible to ground stations for only a short period of time, a working system requires many satellites in orbit, keeping one spacecraft in view at all times. Hence the requirement that the spacecraft be small and economical.

Several companies have announced plans to install global, low-earth orbit satellite communications systems. Notable among these are Motorola, with its Iridium system, Orbital Sciences Corporation of Dulles, Virginia with its Teleglobe System, and Starsys Global Positioning Inc., of Lanham, Maryland.

While initial deployment of these systems will utilize larger rockets which can carry several spacecraft at a time, future satellite replacements will be one at a time, providing a significant market for small launch vehicles.

6-2 Pegasus

The Pegasus space launch vehicle is produced by Orbital Sciences Corporation of Fairfax, Virginia. It was privately developed in a joint venture between Orbital Sciences and Hercules Aerospace Company. Pegasus is the first all-new unmanned space launch vehicle developed in the United States in over twenty years. The vehicle was designed specifically to launch small payloads into low-earth orbit.

The Pegasus concept is unique in that it is launched from a carrier aircraft, in contrast to conventional launches from ground launch pads. A specific advantage of the system is its ability to launch payloads into any inclination by flying the carrier aircraft to an appropriate launch point. No other vehicle can boast of all-azimuth launch capability.

Pegasus can carry a range of payloads from 400 to 1000 pounds in weight. The vehicle is 50 feet long and 50 inches in diameter and has a gross launch weight of 40,000 pounds. It is a three stage device, employing three solid rocket motors, and utilizes a 3-axis inertial guidance system. The vehicle structure is built of graphite composite.

Roll out of the first Pegasus Space Launch Vehicle

Hercules Aerospace Company provides the solid rocket motors for Pegasus. All three stages are new designs and utilize epoxy-composite motor cases.

Pegasus was developed with $45 million of private capital. Development costs were held to a minimum through utilization of components that were already developed. The missile's flight control computer is the fire control computer used on Israel's main battle tank. The guidance system is the Litton produced system for the Navy's Mk. 48 torpedo. The system uses the B-52 carrier plane's guidance system for reference until launch.

The carrier aircraft for early Pegasus flights was NASA Dryden's B-52. The program has since acquired a Lockheed 1011 and modified it to accommodate Pegasus. This gives the Pegasus program independence from other programs for which the B-52 is scheduled.

Orbital Science's first customer for a Pegasus flight was DARPA (now ARPA). The first flight was in 1990, three years after the project was conceived. ARPA plans several more flights for Pegasus. Meanwhile, the company has pressed forward to acquire commercial customers for its system.

Pegasus Air-Launched Space Booster mounted on the wing of NASA Dryden B-52

6-3 Taurus

Orbital Sciences was the winner of a Standard Small Vehicle Procurement initiated by DARPA in 1989. DARPA's objective was to develop a low-cost capability to launch payloads weighing up to 3000 pounds into low-earth orbit. Further, in the interests of lowering cost, the agency required that the contractor demonstrate that the launch vehicle and all its support equipment could be hauled to an essentially bare concrete pad and be erected and ready to receive a payload within five days. Another 72 hours for payload integration and checkout was allowed to prepare the vehicle for liftoff.

Orbital Sciences responded with its Taurus concept, consisting of the basic Pegasus launch vehicle, without its wings, placed atop a MX first stage. The first launch is planned to take place in 1994.

Taurus will be capable of launching from Cape Canaveral, Wallops Island, and Vandenberg Air Force Base.

Taurus' most serious development problem arose from using the MX first stage for its booster. The MX booster normally ignites 200 feet above the ground. On Taurus, the distance is about ten feet , which presents major vibration problems. On typical launchers, such problems are mitigated through use of a flame bucket and water deluge to suppress vibration. Orbital Sciences moved quickly to replace the MX booster with the new Thiokol 120 boost stage, which provides more benign starting conditions.

Based on anticipated success of the DARPA program, Orbital Sciences plans to offer enhanced versions of Taurus to the commercial market. Taurus XL will carry up to 3500 pounds to low-earth orbit and 1,140 pounds to geosynchronous transfer orbit (GTO). This capability will be achieved by increasing the length of the Pegasus stage 1 and stage 2 motors. Further plans are to create the Taurus XL/S by adding strap-on booster rockets. This proposed configuration would be capable of carrying 4300 pounds to low-earth orbit.

Curiously, builders of small rockets are exhibiting ambitions to grow out of that arena, as if unwilling to believe that the small satellite business will be sustained. But it is likely that this will be the area of most rapid growth for future spacecraft.

Orbital Sciences has positioned itself as one of the most aggressive members of the space launch community. The rapid growth of the company, drawn from strong support by DARPA establishment, has signalled to the industry that new methods and processes can indeed be instituted in the space launch business, and would-be survivors in the business must act on this realization.

Taurus launch vehicle on its simple launch pad (OSC)

A rival producer of small launch vehicles is EER Systems of Vienna, Virginia.The Conestoga launch vehicle was originated by astronaut Deke Slayton in a company called Space Services Inc., of Houston Texas. The product was taken over by EER Systems when the company fell on hard times. Conestoga has yet to lift a payload into earth orbit. NASA selected a Conestoga in 1991 to launch its COMET commercial return capsule. The project is presently a victim of government budget constraints.

The Conestoga rocket is a modular design based on using various combinations of the Thiokol Castor IV solid rocket motor, coupled to a selection of solid upper stages. The booster consists of a core Castor IVA or B solid rocket motor surrounded by two to six more Castor IVA orB motors. The upper stage above the core motor may be from the Thiokol STAR 37, 48, or 63 series. Payload capability of the Conestoga ranges from 800 pounds to 3300 pounds to low-earth orbit, depending on the selected modular design.

The vehicle carries a small hydrazine fueled maneuvering and attitude control stage above the STAR motor upper stage to provide velocity augmentation and control of pitch, yaw, and roll.

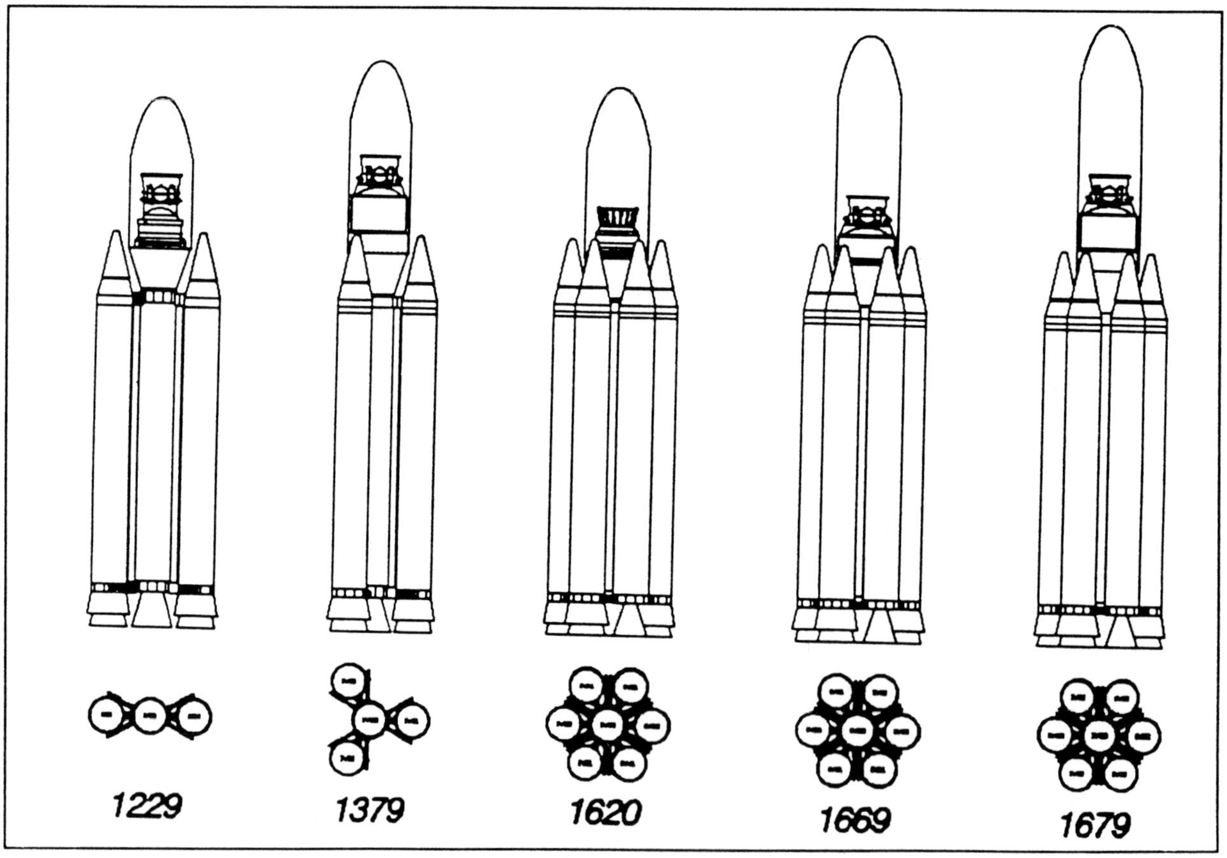

Conestoga line of launch vehicles (EER Systems)

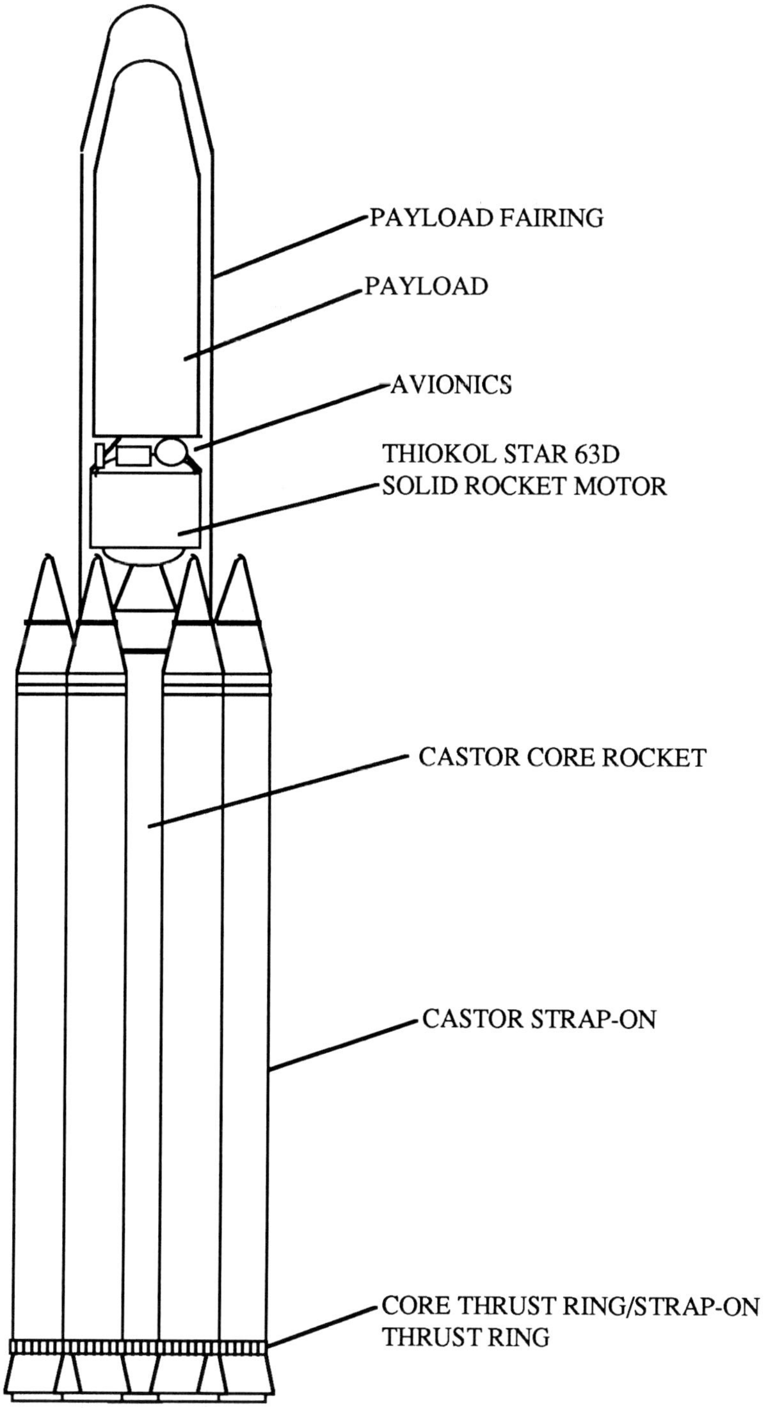

Conestoga launch vehicle as configured for the COMET payload (EER Systems)

6-5 Scout

Scout was the first U.S. launch vehicle to use all solid-fuel rocket stages in its construction. It was developed by the Vought Corporation (now LTV) to serve NASA and DoD requirements for launching orbital and atmospheric reentry research payloads. Its development came at a time when the Air Force was funding solid rocket motor programs for ballistic missile applications. The first flights began in 1960 and have continued to the present. The final flight of this vehicle was in 1994.

Scout has undergone many system improvements over the years, while maintaining the same general configuration. Most of the improvements have been directed toward achieving higher performance. The current model is capable of launching 600 pounds to a 100 nautical mile orbit. The basic vehicle is a four-stage device approximately 75 feet in length.

The first stage of the launch vehicle evolved from the Polaris missile, designated the Algol rocket. Originally built by Aerojet, the stage appeared in 1972 as the Algol IIIA, developed by Chemical Systems Division of United Technologies. The Algol IIIA provides a liftoff thrust of 104,500 pounds to the Scout vehicle.

Scout's second stage was derived from the Army's Sergeant missile. This stage, applied to space usage, was more commonly called the Castor, previously mentioned in other chapters in its role as a strap-on booster for Delta, Atlas, and Titan II. Castor IIA, currently used on Scout, is 20.7 feet long and provides a thrust of 60,100 pounds.

The third stage of Scout is the Antares, currently a product of the Hercules company. This stage has gradually evolved through performance improvements to the Antares IIIA. Its overall length is 11.2 feet with a thrust level of 18,200 pounds.

The fourth stage, designated Altair, was a product of the Vanguard program. It too has undergone a series of performance improvements by the manufacturer, Thiokol Corporation. Altair IIIA, the current model, has a thrust level of 5,800 pounds.

Scouts were launched from a facility at Wallops Island off the Atlantic Coast, from Vandenberg Air Force Base in California, and from the San Marco Equatorial Mobile Range off the east coast of Kenya. Operations from the San Marco site have been conducted in conjunction with the Italian Government.

The Scout launch system is no longer in production for U.S. launchings, having performed its final launch in 1994.

Four stage Scout launch vehicle at liftoff (Loral)

6-6 Lockheed's Entry

The Lockheed Company has recently announced plans to enter the small launch vehicle market with its own designs. Lockheed's experience with Polaris, Poseiden, and Trident solid propellant sea-launched ballistic missiles provide an easy entry into the competition for launching small payloads.

The company originally considered conversion of submarine launched vehicles to this purpose but eventually moved to solid stages being made available to the commercial market.

Lockheed plans to produce a family of solid propellant launch vehicles with payloads ranging from 2300 to 9000 pounds to low-earth orbit. The initial market for the new launch systems is believed to be small military and NASA payloads.

The launch vehicle family consists of three models. The smallest vehicle, named the Lockheed Launch Vehicle 1 (LLV1), uses the newly developed Thiokol Castor 120 solid propellant motor as its first stage and the Orbus 21D motor from United Technologies as its second stage. The first flight of this launch vehicle is scheduled for November, 1994. It will have a payload lift capability of 2300 pounds to a 100 nautical-mile orbit. Cost of the launcher is estimated at $14 million.

Lockheed's LLV2 will have a payload of 5300 pounds to low-earth orbit. This vehicle retains the Thiokol Castor 120 solid motor as its first stage and adds another Castor 120 with a modified expansion nozzle as its second stage. The Orbus 21D functions as a third stage.

A further growth version is designated the LLV3. The LLV3 is equipped with from two to six Castor 4 strap-on rocket motors for boost augmentation. This launch vehicle is projected to have a lift capability nearing 9000 pounds to low-earth orbit, which is in the range of the McDonnell-Douglas Delta rocket. Its cost is pegged at somewhat above $20 million, according to company officials.

Ground operations, including vehicle checkout and launch, are intended to be accomplished by a crew of no more than 25 people. All three companies that have announced small launch vehicle programs intend to operate in this manner. A significant cost issue for existing launch systems has been the large "standing armies" required for processing at the Vandenberg and Florida launch complexes. The small launch vehicle contractors intend to remedy this with streamlined and simplified operations.

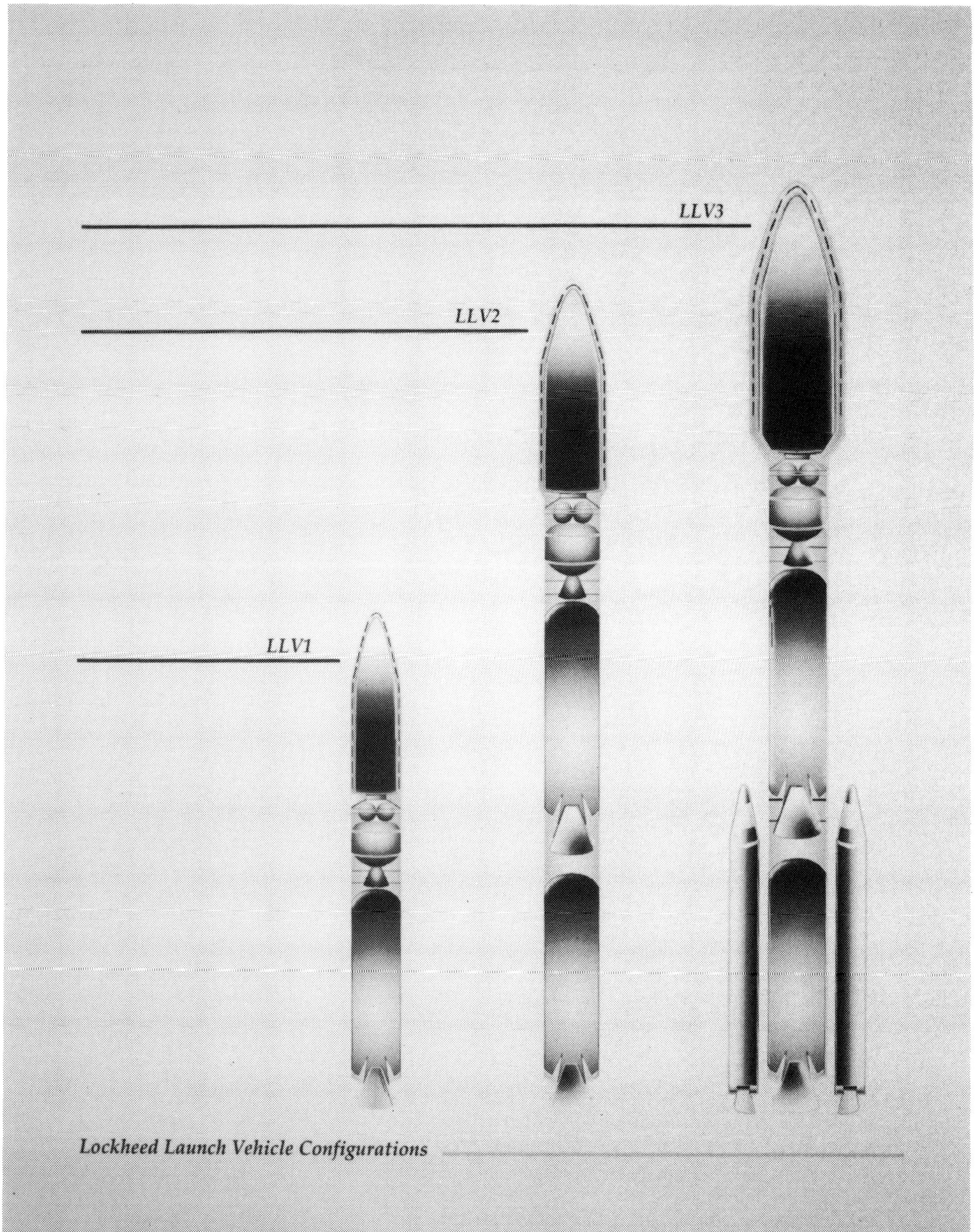

The Lockheed family of solid propellant space launch vehicles (Lockheed)

6-7 Outlook

ARPA's successful development and promotion of small satellites has opened up a new segment for the spacecraft industry that is certain to grow. Spacecraft are electronic devices, serviced by power, attitude control rockets, thermal control, and supporting structure. Miniaturizing the electronics has the effect of reducing requirements for power, thermal control, rocket propellants, and structure. Lower costs for spacecraft are a direct result, so it is only natural that the industry would move rapidly in this direction. The development has made spacecraft affordable for many applications that would previously have been out of the question, had it been necessary to use conventional spacecraft. A project like the proposed Iridium mobile communications system would have been too expensive.

The Department of Defense and NASA can be expected to move rapidly in the direction of small satellite usage due simply to a need to save money. But the market for small spacecraft is not confined to U.S. military, civilian, and commercial arenas. Development of small economical spacecraft has made it possible for many nations around the world to begin to use spacecraft for their own scientific research and civilian service objectives.

The outlook for small payload launch vehicle manufacturers is therefore optimistic, which explains the aggressive entry by Lockheed into a field where there are already a half dozen other players. It is likely that other manufacturers will be also lured by the business prospects, and will develop their own launch vehicles. A highly competitive atmosphere will result, following which one can expect a shakeout, leaving not more than two manufacturers with the major part of the business. At the moment, Orbital Sciences and Lockheed appear to hold strong positions. Orbital Sciences, because it already has a backlog of launch contracts, and Lockheed, because it has both financial clout and established rapport with military and NASA customers.

Manufacturers of small launch vehicles can anticipate a long-term future for this business. Replacement satellites for the mobile satellite communications systems alone will provide a bread and butter business base.

Lockheed's venture into small payload launch vehicle production has taken an interesting turn with the announcement of a merger between Lockheed and Martin Marietta on August 29, 1994. The new company, Lockheed-Martin, could elect to continue development of the solid rocket family as part of its extended launch vehicle product line, or it could shelve it, as having insufficient potential for future business.

7

Solid Rocket Boosters

7-1 A Bit of History

Although solid propellant rockets can be traced back to early Chinese history, they first appeared in the western world with the invention of the war rocket by Sir William Congreve during the early 1800's. The rockets were first used in a British attack on Boulogne in 1806, with considerable success. The devices were fired from rocket tubes mounted on tripods. Sir William also invented a rocket for killing whales.

The Congreve rocket was superseded by the Hale rocket. This device consisted of a hollow head filled with explosive mounted on a cylindrical steel case containing the rocket composition. The burning propellants were expelled through a single aft port over a screw that was attached to the base of the rocket. Impingement of the exhaust gases on the screw provided the rotational motion needed for stability during flight. The Hale rocket was launched from a trough. The range of these rockets was a scant 1200 yards and the accuracy was poor.

These engines of war were used in the British service until about 1885. Their use by armies on the continent was discontinued about twenty years earlier. Until the mid-20th century, the use of solid rockets was confined to fireworks and signal devices. It was only after World War II that serious development of large, solid propellant rockets was begun to meet the requirements of the U.S. IRBM and ICBM programs.

A number of companies moved into the business and the technology advanced rapidly. Lockheed, Aerojet, Hercules, Thiokol and United Technologies emerged as the principal players in this arm of rocketry. While the technology was directed primarily at development of multi-stage vehicles for launching out of submarines and land-based silos, there was no shortage of ideas about the potential of solid rockets for space launchers.

In one of the more ambitious ventures during the Apollo years, Aerojet fabricated a solid motor 260 inches in diameter at a Florida facility. The propellant was cast into a case that was emplaced vertically in a hole in the ground. The rocket was test fired successfully from that position. This was the largest solid rocket motor ever built.

Application of solid-propellant rockets to space launch vehicles finally appeared in the 1960's with the development of the Titan III rocket. Titan III was developed by the Air Force for the Dyna-Soar project. The rocket was built about the Titan II core vehicle, increasing performance greatly by adding two solid-propellant strap-on rockets.

7-2 Shuttle Boosters

Two solid rocket boosters provide the main thrust to the Space Shuttle, beginning at liftoff and continuing until the shuttle reaches an altitude of about 150,000 feet. The boosters provide the main support for the shuttle and the external tank as the assembled vehicle sits on the launch pad. The orbiter is attached to the external tank and the external tank is attached to the solid rocket motors. The motors are the largest ever flown and are designed to be reused.

The motor casings are produced by the McDonnell Douglas Astronautics Company of Huntington Beach, California, and the propellant segments are produced by Thiokol Chemical Company in Brigham City, Utah. The complete motor is too heavy to ship, so the propellant segments are shipped separately by rail to the Florida launch site where they are assembled, along with other key components, to create the flight-ready motor. The main parts of the rocket motor are the forward dome, six propellant sections, an aft dome, an aft attachment ring for the external tank, and a pair of stiffener sections.

A single Shuttle solid rocket booster, or SRB, is 149 feet long and 12.17 feet in diameter. A complete SRB assembly weighs 1,256,000 pounds and produces 3,310,000 pounds of thrust at high altitude.

Three major subassemblies make up the complete SRB. They are the nose cone, which contains avionics and recovery equipment, the motor assembly described above, and the exhaust nozzle assembly.

The SRB nozzle, which weighs over ten tons, converges from its attachment to the solid motor to its throat, then diverges again to form the exhaust port. A thick carbon-cloth liner protects the outer metal structure from high exhaust temperatures. The liner erodes and chars during the burn process.

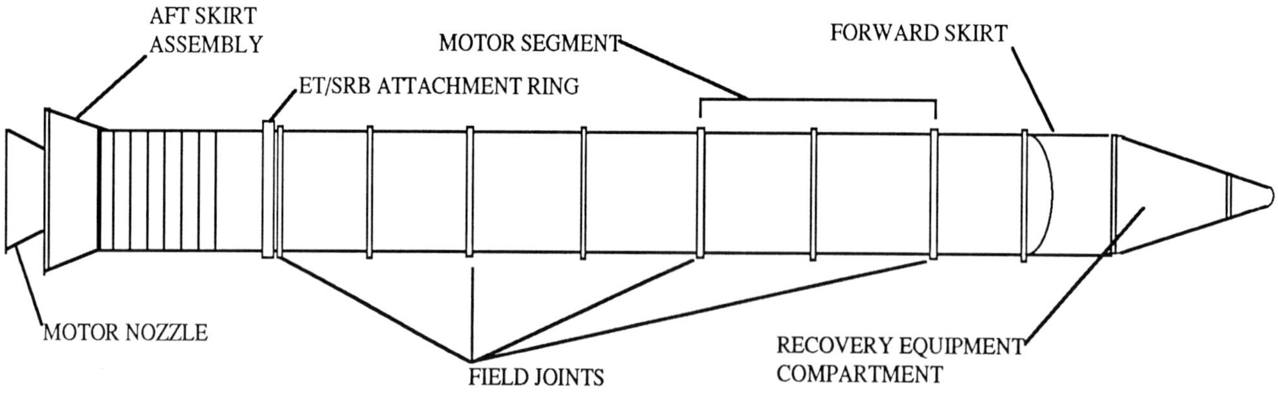

The Space Shuttle Solid Rocket Booster (NASA)

The divergent part of the nozzle is mounted on a bearing that permits swivelling, to provide steering for the Space Shuttle. The nozzle can be swung up to eight degrees by means of two large hydraulic actuators that are affixed to it.

The SRB's are released from the Shuttle following burnout by means of pyrotechnic devices at their attachment struts. Lateral firing solid rockets then move them away from the Orbiter, which remains attached to the external tank. The SRB's reach an altitude of about 42 miles before falling into the ocean approximately 140 miles from the launch site. During descent, parachutes are deployed from the nose cone to reduce the velocity at water entry, and to ensure a tail-first landing. A recovery ship intercepts the SRB's and tows them back to port for refurbishment and re-use.

As is expected in programs of this nature, Space Shuttle managers recognized that the SRB could be improved for better safety, lower cost, and greater performance. The Space Shuttle Challenger accident and subsequent stand-down forced the adoption of safety improvements but was also instrumental in Congress directingNASA to undertake development of an advanced solid rocket motor (ASRM). The new design was to take advantage of fifteen years of advancing technology since the SRB was originally designed. It was also to improve the performance of the shuttle to offset the penalties incurred by the safety improvements, as well as to provide the ability to carry full payloads to the higher altitudes slated for space station Freedom.

In the ensuing competition a Lockheed-Aerojet team emerged the winners. Plans call for production and pouring of the motor propellants at a new, highly automated factory at Yellow Creek, Mississippi. The contractor for the motor casings is Babcock & Wilcox of Mt. Vernon, Indiana. The ASRM propellant was planned to be mixed and poured in a continuous mix process with in-line quality control. Continuous mix processes are replacing batch mix processes in the solid propellant industry, offering the dual benefits of lower cost of production as well as improved safety. Better safety comes about as a consequence of the small quantity of propellant that is moved through the continuous mix process as compared to the large quantities in the batch process.

The ASRM booster is four inches greater in diameter than its predecessor, but is the same length. The new booster weighs 1,344,600 pounds and generates 3,485,000 pounds of thrust at high altitude. The improved performance of the ASRM is expected to restore the Space Shuttle's capability to carry its specification payload of 65,000 pounds to low-earth orbit.

Space Shuttle safety and reliability are expected to be improved with the introduction of the new motor. A number of primary catastrophic failure modes have been removed in the new design. Several improvements were made in the motor case, reducing the number of joints and eliminating potential for further failures of the type that caused the Challenger accident.

The ASRM contains a new propellant grain design which allows tailoring the thrust at peak aerodynamic load to a sufficiently low level that throttling of the Orbiter's main engines at this point in flight will no longer be necessary. Elimination of this operation is believed to increase the reliability of these engines. The solid propellant composition is also different, yielding slightly improved performance over the SRB.

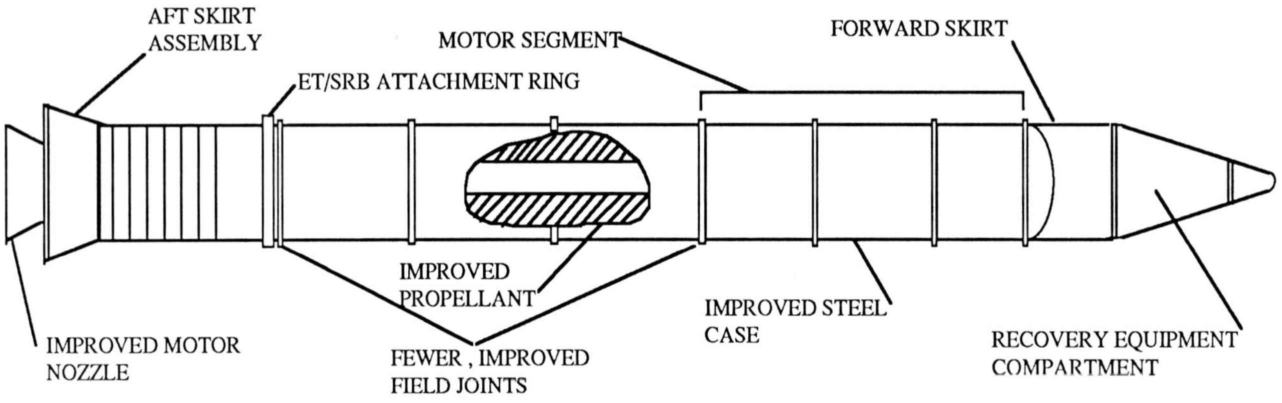

The Space Shuttle Advanced Solid Rocket Motor

The ASRM features an improved ignition system and a new nozzle design. The new nozzle weighs approximately 5,000 pounds less than its predecessor. The new nozzle has fewer joints, eliminates 72 seals, and includes an erosion resistant carbon-carbon wall at the throat entrance.

NASA planned to test the ASRM extensively at the Stennis Space Center in Mississippi. The program would duplicate the test process that was imposed on the SRB after the Challenger accident. Additional testing would verify that no new difficulties will be experienced with the assembly, installation, and operation of the new motor.

The ASRM began as a program costing somewhat over $1 billion . First flights were to occur in 1996. However, severe cost escalation since the program started led to refusal by Congress to provide further funding when cost to complete was pegged at over $3.5 billion. The program was cancelled in 1993.

7-3 Titan Boosters

Titan solid rocket boosters date back to 1957, shortly after Sputnik I flew, when the Air Force developed preliminary requirements for a manned, reusable spacecraft. Project Dyna-Soar, as previously discussed in Chapter 4, evolved from these requirements.

Launching of the new space plane required a lift vehicle with a capability far greater than either Atlas or Titan II could provide. The higher lift capability was achieved with a new launch vehicle, Titan III, which used the basic Titan II with two strap-on solid rocket boost motors. The Dyna-Soar project was cancelled in 1963 for lack of a definite mission in the Air Force and absence of interest by NASA, who had other plans for achieving manned space flight.

But the Air Force came away from Dyna-Soar with a new launch vehicle having a lift capability of over 30,000 pounds to low-earth orbit. Titan III, flown for the first time in 1965, enabled the Air Force to undertake deployment of large surveillance spacecraft. Over the next twenty years, Titan III and its uprated versions served the nation well for Air Force and NASA missions. One version of Titan III, identified as Titan IIIE, was equipped with a Centaur upper stage . It was used by NASA for launching the spectacularly successful Viking and Voyager planetary missions.

The Titan III solid rocket booster is built by Chemical Systems Division of United Technologies, Inc. The ten-foot diameter motor consists of five and one-half segments which are bolted together after shipment to eastern or western launch complexes. Titan III boosters are almost out of production, their single market presently being the commercial Titan III, which itself is on hold.

When Martin Marietta company undertook to develop Titan IV during the mid 1980's, one objective was to improve the performance of the Titan 34 D version of Titan III. This was achieved by "stretching" the core stage and replacing the five and one-half segment rockets with a seven segment version, raising the payload capability of the system to 39,000 pounds to low-earth orbit. The seven segment rockets are also built by Chemical Systems Division.

The two strap-on boosters weigh 300 tons each and measure 112 feet in height. Thrust of each motor is 1.6 million pounds.

Thrust vector control on Titan boosters is achieved by injecting fluid into the exhaust nozzles, thus deflecting the exhaust stream. The nozzles are ringed with valves which control the injection zone. The fluid is nitrogen tetroxide, the same as the core stage oxidizer, and is supplied from a pressurized tank attached to each booster.

Five Segment Booster Rockets on a Commercial Titan (Martin)

In a program designed to raise the payload capability of Titan IV to 50,000 pounds to low-earth orbit, the Air Force has undertaken development of a new booster motor under the designation "Solid Rocket Motor Upgrade (SRMU)." The contract for development was awarded to Hercules Aerospace Corp. of Magma, Utah. The work is being done under subcontract to Martin Marietta.

Titan IV launch showing seven-segment solid propellant rocket boosters (Martin Marietta)

Thiokol developed the Castor strap-on rocket motor for the Delta launch vehicle, which is produced by McDonnell Douglas Aerospace Company. For many years the Delta rocket used nine Castor IVA strap-on boosters atttached to its single-engine liquid-propellant core stage. As demand for this class of boosters spread to other launch vehicles, it was only natural that other manufacturers would enter the field.

The Castor IVA motor has a steel casing and a stationary exhaust nozzle. The exhaust nozzle is canted eleven degrees so that it points away from the core vehicle in its strap-on postion. Castor IVA is 40 inches in diameter and 362 inches long. It weighs 25,562 pounds and develops an average thrust of 110,430 pounds.

Castor IVA-XL is the last of a line of Castor motors developed by Thiokol Corporation. The motor is 40 inches in diameter, and 460 inches long. It weighs 32,500 pounds, and has an average thrust of 138,000 pounds. Its performance is about 30% greater than the Castor IVA and was developed when Hercules Aerospace developed a competing motor, wresting the Delta business away from Thiokol. The Hercules GEM motor is similar in geometry and performance but uses a graphite-composite motor case for a weight advantage over the Castor IVA-XL.

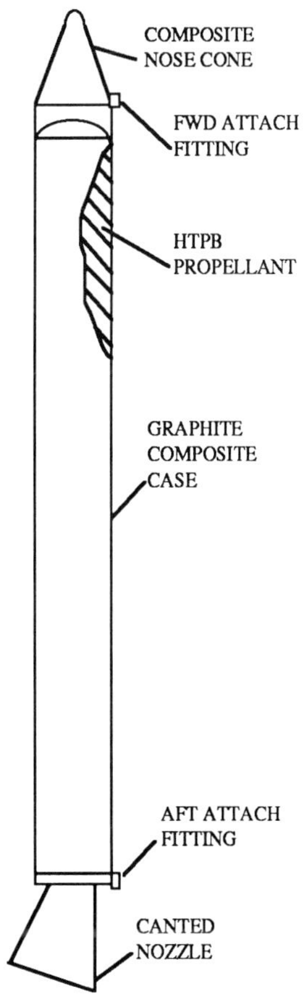

Hercules GEM solid motor

General Dynamics Space Systems Division has adapted the Castor IVA to its Atlas II launch vehicle in a modification designed to raise its lift capability. The Atlas II AS model uses four strap-on rockets which are ignited shortly after liftoff. Future uprated versions of the Atlas IIAS could use the Castor IVA-XL or the Hercules GEM motor.

Martin Marietta has adapted the Castor IVA to its Titan II launch vehicle to improve its lift capability to low-earth orbit. As many as nine Castor IVA's can be used on each flight. Castor IVA-XL or GEM motors are likely for uprated versions.

Lockheed Aerospace plans to use Castor type booster rockets on its LLV 3 small payload launch vehicle, the largest of three launch vehicles the company is developing with private funding in its bid for the growing small payload market.

EER Systems' Conestoga launch vehicle is made up of Castor motors in each of its several stages. It uses a cluster of four Castor motors in its first stage, three motors in its second stage, and one in its third stage. Upper stages have modified exhaust nozzles for operation at high altitude. EER Systems has followed the lead of the Scout launch vehicle in applying Castor rockets to upper stage functions.

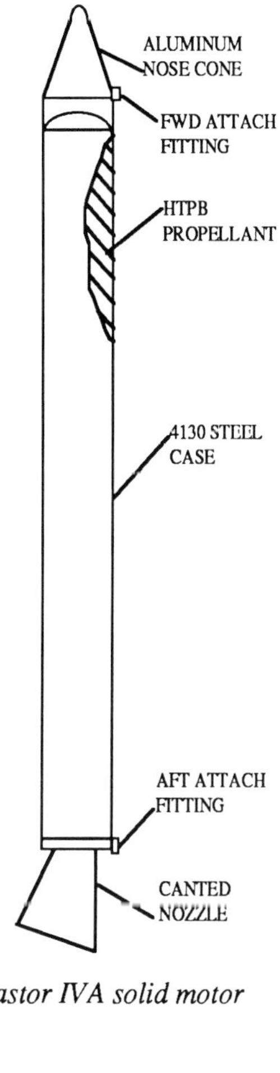

Thiokol Castor IVA solid motor

Emergence of the small payload industry as a result of ARPA's small payload development program has opened the field for small payload launch vehicles. Thiokol has responded to a perceived need for a simple solid rocket motor in this arena by developing its new Castor 120 motor.

The Castor 120 motor is designed to be used as a booster rocket or as a second stage in small launch vehicle applications. It can also be used as a strap-on booster for larger payload launch vehicles. The thrust profiles can be readily tailored to the requirements of different launch vehicles by machining alternative propellant grain configurations.

The motor is 93 inches in diameter, 426 inches long, and weighs 118,000 pounds. It produces an average thrust of 386,000 pounds.

The new motor has already been adopted by Lockheed for the first stage of its LLV1 launch vehicle and for the first and second stages of its LLV2 and LLV3 vehicles. It has also been selected by Orbital Systems as the booster for that company's Taurus launch vehicle.

7-5 The Surplus Inventory

The end of the Cold War and dismantling of the nuclear missile fleets will release a large number of solid propellant rocket stages that could either be destroyed, or be adapted to use for small payload booster and upper stages. The inventory is viewed by some as a source of cheap stages whose use could result in large savings in launch costs. It is viewed by others as a threat to a growing private industry that is making strong efforts to provide economical launch systems.

The Department of Defense has opted to proceed slowly in any program that would release surplused stages to the industry. Questions regarding safety of war rockets, particularly in those that use sensitive propellant combinations enter into the concerns that are being raised. Age of the stages raises questions regarding the reliability that can be expected. Overall, it is not at all clear that use of surplus rocket stages save money.

Decisions by Orbital Sciences to use the Thiokol 120 for its Taurus launch vehicle, and by Lockheed to use the 120 for its small launch vehicle family indicates that industry managers have already performed trade studies concluding that new, commercially produced rocket motors are best suited to their purposes.

Surplus rockets will eventually be put to good use, but the likely applications will be within the DoD's own research programs, using launch facilities that were originally built for development testing and flight certification.

Castor 120 motor in a Thiokol test bay prior to test firing, April 22, 1992 (Thiokol)

7-6 Outlook

The Titan IV SRMU has had a troubled development history which has included one accident and severe cost overruns.These events have led to skepticism on the part of some members of Congress regarding the worth of the program. By Congressional direction, Martin Marietta and its subcontractor Hercules Aerospace are to complete development tests before building production articles. These tests were scheduled for completion in 1993. Current plans call for production of fifteen pairs of boosters over a period of four years rather than one year as originally scheduled.

Slowdown of the Titan IV program to three or four flights a year has a dimming effect on the future of the SRMU, a part of which rests on the Air Force's declining needs for payload carrying capability in excess of what can be done with the current Titan IV. Additional problems during development could result in the program's cancellation. As it stands, SRMU has not been a profitable venture for Hercules Aerospace. Prospects for a robust production program are slim.The SMRU could enter an extended production hiatus after the initial production run.

Development work on the Space Shuttle ASRM ceased when Congress directed cancellation of the program in October , 1993. Work on a new booster for the Space Shuttle could be initiated again, but it is more likely that the new booster will be liquid fueled. NASA officials want the extra performance afforded by new boosters, arguing that fewer Shuttle flights will be required to build the Space Station, and relocation of the station orbit to an inclination better suited for joint Russian and U.S. operations will require booster rockets with better performance.

The future for small solid rocket boosters is quite bright.The rapidly emerging market for small satellites and simplicity of the solid rocket vehicles designed to launch them into orbit forecasts a good business for industries that are positioned to work this end of the launch vehicle spectrum.

8

Delta Clipper Single-Stage-to-Orbit

8-1 The Dream

Since the beginning of modern rocketry, the challenge for rocket design has been the simplest of designs: a rocket that could enter earth orbit using a single stage, carrying a useful payload. But the limits of rocket engine performance, coupled with the inert mass that makes up the launch vehicle structure, propulsion, avionics, and other large components were always such that a single stage could not attain the required velocity to reach earth orbit while carrying a payload weight large enough to be of interest.

An engineering team at McDonnell Douglas Space Systems Company believes that it has found the answer. Moreover, the concept being promoted goes the single-stage-to-orbit (SSTO) one better. After depositing its payload into low-earth orbit, it can be returned to earth and gently landed by means of its rocket motors at any appropriate landing spot. The McDonnell Douglas reusable SSTO is dubbed the Delta Clipper.

In August of 1991, the Strategic Defense Initiative Office awarded a $62 million technology demonstration contract to McDonnell Douglas. The contract award resulted from evaluation of several competitive proposals. The contract requirement was to design a reusable SSTO that could carry large numbers of small payloads into earth orbit. The vehicle was to carry up to 20,000 pounds of payload on each flight at a cost of less than $1000 per pound. This compares to costs ranging from $5000 to $10,000 for current expendable launch vehicles.

The purported attractiveness of reusable launch vehicles is a low cost of operation. It stands to reason that a reusable device should be cheaper than one that is thrown away after each flight. But the astute observer, with only a minimum knowledge of space transportation history, will be quick to question this. The Space Shuttle was built on the same premise. It turned out to be the most expensive launch system ever built. One can argue that any new reusable system has to demonstrate that it can not only close the gap in costs between high cost reusable systems and expendable launch vehicles, but also has to demonstrate that this time it will produce great savings. A formidable challenge.

McDonnell Douglas believes that it can meet the performance and cost challenges in a design concept that minimizes the launch vehicle weight through the use of advanced materials, uses liquid hydrogen and oxygen for rocket propellants, and is equipped with new rocket engines that will perform efficiently at all altitudes.

8-2 Delta Clipper Concept

The flight version of the Delta Clipper (DC-Y) will stand 127 feet high and will be 30 feet wide at the base. It will weigh approximately 80,000 pounds empty and will burn 940,000 pounds of propellant in each mission. It can carry 20,000 pounds of cargo to low-earth orbit.

A rocket engine design for the vehicle has not yet been selected. The choice lies between use of multiple engines with bell nozzles, optimized to operate efficiently at all altitudes, or some variation of "aerospike" or "plug nozzle" designs, which are altitude compensating as a result of their basic geometry. The latter have not yet been demonstrated successfully in a complete engine by U.S. engine companies.

The DC-Y external geometry is determined by the selected mode of atmospheric reentry. Unlike other SSTO designs, which enter the atmosphere base-first, the DC-Y is designed to enter nose-first at a high angle of attack, averting the design problems associated with protecting the rocket engines from reentry heating. To do this, the vehicle has a lifting body shape. Using extendible spoilers at the base of the vehicle, it has a projected cross-range maneuvering capability approaching 1600 miles.

Artist's concept of the Delta Clipper SSTO during launch (McDonnell Douglas)

The DC-Y design draws heavily on recent advancements in materials for its structure and fuel tanks. Research work on the National Aerospace Plane, the National Launch System, and other projects has produced light weight materials that can handle the difficult thermostructural environments that would be encountered in proposed modern launch systems.

The structural framework of the Delta Clipper is made of graphite epoxy composite. Overlaying the framework is the shell, or skin, which is made of a graphite /epoxy and aluminum in a honeycomb sandwich construction.Thermal protection during flight, particularly for the severe reentry conditions, comprises an exterior carbon-carbon and carbon-silicon carbide insulation. The landing gear is made from titanium-silicon carbide composite.

The propellant tanks are suspended inside the vehicle's structural frame. The larger tank, containing liquid hydrogen, is positioned aft and is made of graphite composite. The smaller tank, containing liquid oxygen, is positioned forward and is made of a new aluminum-copper-lithium alloy.

If conventional bell nozzle rocket engines are employed on the Delta Clipper, the aft end of the vehicle will be a blunt shape with the engine nozzles protruding through openings in the shell. The engines will be mounted on a structure that will be a part of the vehicle's frame. For best performance at high altitude, extendible nozzles may be used.

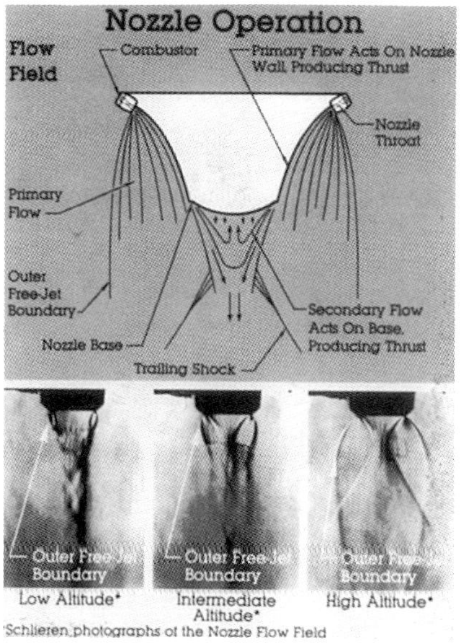

Aerospike engine configuration (Rocketdyne)

If an aerospike engine is developed for the vehicle, the aft end takes on an entirely different shape. An aerospike engine will feature a toroidal combustion chamber, or many small chambers that will encircle the vehicle. The chambers' exhaust gases expand against a large central cone or plug, which is either high temperature resistant or cooled with hydrogen fuel. The turbopumps which feed propellants to the combustion chamber would be mounted on a structural framework inside the exhaust cone. Large questions remain regarding the performance of aerospike engines at all altitudes, as well as thermostructural weight penalties associated with a circumferential engine that approximates 90 feet in length.

Delta Clipper's payload will be carried in compartments just aft of the vehicle's nose. Doors in the wall of the vehicle provide for cargo ground installation and deployment of payloads when the vehicle reaches orbit.

Delta Clipper's return flight plan calls for employment of maneuvering thrusters and deployment of four extendible control surfaces around the base of the vehicle. When the craft has reached 20,000 feet altitude and 180 knots, the engines ignite in an idle mode and the vehicle is put in a vertical position. The craft then hovers, descends, and lands.

8-3 Delta Clipper Demonstrator

McDonnell Douglas completed fabrication of a one-third scale demonstrator vehicle, designated the DC-X, and accomplished limited testing during 1993. The DC-X is not capable of orbital flight, but instead was intended to verify some of the hardware configurations and functions anticipated in the future operational vehicle. These efforts are a continuation of the Strategic Defense Initiative contract under which the SSTO development was initiated. Work on the Delta Clipper is not funded by the government, but continues under independent research and development at McDonnell Douglas Space System Division.

Pratt & Whitney low altitude engine developed for the DC-X

After completing fabrication of the DC-X at the Huntington Beach factory, the development team moved the demonstrator to the White Sands Test Facility in New Mexico to perform evaluation tests.

Initial tests consisted of static firings to evaluate the propulsion system, including four Pratt & Whitney RL10-5 engines that were developed for the demonstrator.

Static firing tests of ten seconds duration verified environmental conditions within the vehicle's engine compartment. A 50 second firing then simulated the profile for DC-X's first hover flight tests. A final firing simulated the most demanding DC-X mission, which called for vertical takeoff, ascent flight, rotation maneuvers, and vertical landing.

Flight tests were conducted at the White Sands Missile Range Space Harbor area. The DC-X was mounted on a launch pedestal about five feet off the ground.

In the first three flight tests the DC-X climbed successively to altitudes of 150 feet, 300 feet, and 1200 feet, landing again on a pad a few hundred feet from the launch pad. With the completion of the last test on September 30, 1993, the program ran out of money. Two additional scheduled flight tests had to be cancelled. McDonnell Douglas hopes for appropriation of $5 million to complete the test program. Additional funding to initiate the next phase of the program under ARPA's directionwas appropriated in the 1994 budget but the Air Force appears to be reluctant to proceed.

The DC-X tests are less a technology development than a hardware demonstration. Maneuvers of the types to be performed were operational practice for the Apollo astronauts over twenty years ago when they repeatedly landed and took off from the moon. Since none of the hardware built for the demonstrator will find its way into an actual flight article, the results of the tests will likely be of little importance to future development of the Delta Clipper concept.

DC-X

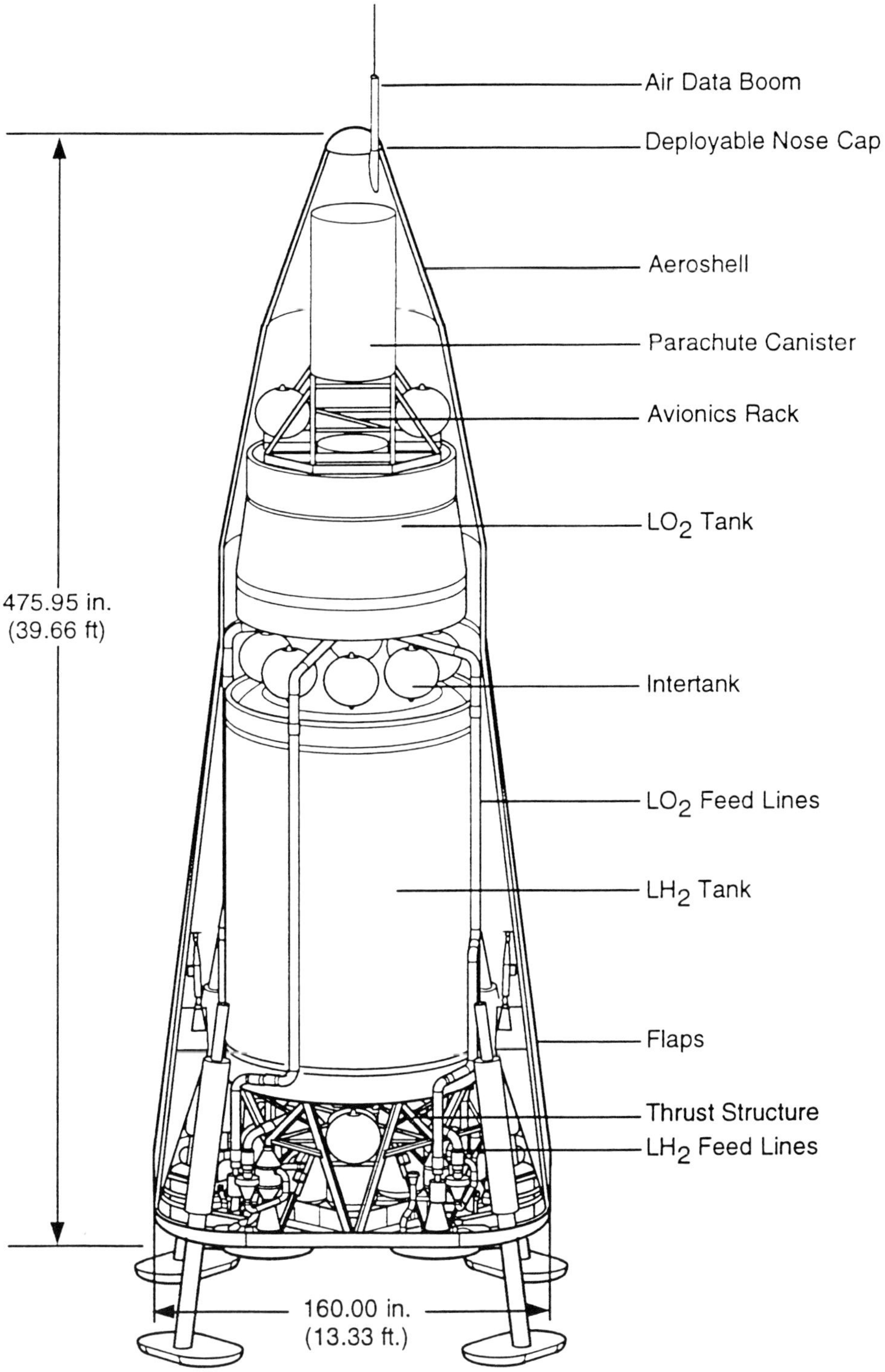

The DC-X demonstrator designed for initial SSTO testing (McDonnell Douglas)

8-4 A Vision in Peril

Considerable skepticism surrounds the Delta Clipper concept in the space engineering community, although the idea is wildly popular among space supporters and has several avid congressional supporters.

The problem reverts to the same issue that has vexed designers in the past. The payload that is projected to be carried by the Delta Clipper is such a small percentage of the vehicles' gross weight, that errors in estimating the weight of the initial design can cut severely into or even eliminate the payload capability by the time the design is complete. Little omissions can add weight rapidly. A simple example is the recent discovery on the DC-X that vortex shedding can be a severe problem for the vehicle sitting on a launch pad, which necessitated the addition of strakes around the nose of the vehicle to alleviate the condition.

McDonnell Douglas' design approach to minimize launch vehicle weight by adopting advanced materials is realistic, but it does not do much more than compensate for having to build a four-layer vehicle that constitutes the tanks with their insulation, a structural frame, a skin, and exterior thermal insulation. It would be difficult enough with this design to build a vehicle that would simply go to orbit and deposit a payload. But the Delta Clipper must go into orbit and carry sufficient excess propellant to ensure a safe return and landing. It must also be sufficiently robust to withstand repeated usage under extreme vibration and thermal environments.

By multistage standards, Delta Clipper at a million pounds gross weight is a very large vehicle for carrying a 20,000 pound payload into orbit. The Atlas IIAS has that capability at half the launch weight. But it is interesting that the concept sits at the lower extreme of a size regime over which single-stage-to-orbit vehicles do become practical. In Chapter 12, a discussion of heavy lifters emphasizes the practicality of single-stage-to-orbit for large payloads.

Single-stage-to-orbit launch vehicles work in large sizes not because designers use advanced materials, but because the vehicle's propellant containment efficiency, or mass fraction, rises rapidly as a function of increase in the gross weight. The weight of propellant rises roughly as a cube function while the inert weight of the vehicle rises as the square. Should the Delta Clipper fail to meet it performance predictions during future work, the designers need simply to make it bigger, and then it will work.

When and if the program moves toward actual development of the Delta Clipper, it would be prudent for its sponsors to require a rigorous independent analysis to verify McDonnell Douglas performance predictions.

While the Delta Clipper may falter by reason of its drive to achieve success through maximizing performance of materials and propulsion, other ways should be seriously examined before the idea is once again abandoned. After all, the idea of a returnable single-stage-to-orbit is fundamentally attractive, and may in the long run prove to be the accepted mode of transportation to space. But one of the prime requisites for such a vehicle is a robustness that will ensure its reusability over many cycles and require a minimum of maintenance between flights. If hardware has to undergo rigorous testing after every flight, and significant replacement, as is the case on the Space Shuttle, then prospects for adoption of this transportation scheme are dim indeed.

An approach that leads to the desired robustness will doubtless center on achieving better propulsion efficiency. Chemical propulsion, in which the launch vehicle carries all of its propellant, reaches its practical limits with hydrogen fueled engines, which yield the best performance. Rocket engine manufacturers already know how to build robust engines of moderately high performance, and engines of lesser durability that perform at somewhat higher levels. A prudent design approach would seek ways to avoid requiring the highest possible performance, preferring engine robustness instead.

A large jump in capability can be acquired if external air is used for oxidizer instead of on-board liquid oxygen during the period of flight through the atmosphere. Methods of doing this may involve new engine designs where a part of the cycle is all-rocket propulsion and a part is air augmented. Another way is to use separate engines which are used at appropriate segments of the flight ascent. For example, a vehicle could be configured to lift off with rocket power, switch to ramjet mode when a sufficient velocity is reached, switch again to scramjet at an appropriate velocity, and then revert to rocket power once more to accelerate the vehicle into orbit. Such a vehicle would be required to carry only a small quantity of oxygen in on-board tanks, resulting in startling improvements in performance.

Rocket engine manufacturers have already performed studies of hybrid engines of the type discussed above. Competing proposals for the SSTO submitted to the Strategic Defense Office in 1989 included their concepts, based on perceived requirements for more efficient propulsion than could be afforded by conventional liquid rocket systems.

Significant research on scramjet engines, conducted under the NASP project, as well as experimental work in Russia, and England, lead to suggesting this type of device as a possible way to provide thrust augmentation for a SSTO. SSTO geometrical symmetry lends itself to adoption of the simplest conceivable scramjet, so development cost can be minimized.

The accompanying sketch shows a concept for a small SSTO that is capable of lifting several thousand pounds to low-earth orbit. Its rocket propulsion system consists of four RL-10 hydrogen-fueled engines, fitted with extendible nozzles to obtain good performance at liftoff as well as at high altitude.

The vehicle is somewhat similar in appearance to the Delta Clipper, except for the specific lifting body shaping. This vehicle would reenter the atmosphere blunt end first, and is provisioned with closures to protect the rocket engines during reentry.

The scramjet thrust assist engine consists of a combustion chamber located at the vehicle's center line (scramjets like long combustion chambers). The walls of the chamber are hydrogen cooled. At the forward end of the tube is a movable plug to adjust entry conditions as the craft moves through the scramjet's operable velocity range.

The propellant tanks are toroidal in shape, surrounding the central chamber. They are attached to the vehicle structure by means of suspension links. Both tanks are foam insulated.

The vehicle structure is aluminum, with a carbon-carbon cover layer in regions of high heating. The blunt aft face is moveable, to enable closure of the engine ports.

The SSTO rests on five retractable landing legs, which can compensate for uneven landing areas.

The cargo compartment opens through side doors into a volume just aft of the vehicle's nose.

During its operating cycle, the SSTO lifts off under rocket power and accelerates to a velocity that enables takeover by scramjet. The vehicle then accelerates by scramjet until it can no longer operate efficiently in that mode. Rocket engines then propel the craft to orbit where it deploys its payload. The main engines are then fired to provide thrust for reentry. The rocket engines are fired again when the craft approaches its landing zone.

A launch vehicle built along these lines not only has future potential for transporting payloads to low-earth orbit, but also has potential for ballistic point-to-point transport from any serviced landing point to any other such point on the surface of the earth. It would be able to do what the National Aero-space plane could do, but in an alternative and possibly simpler fashion.

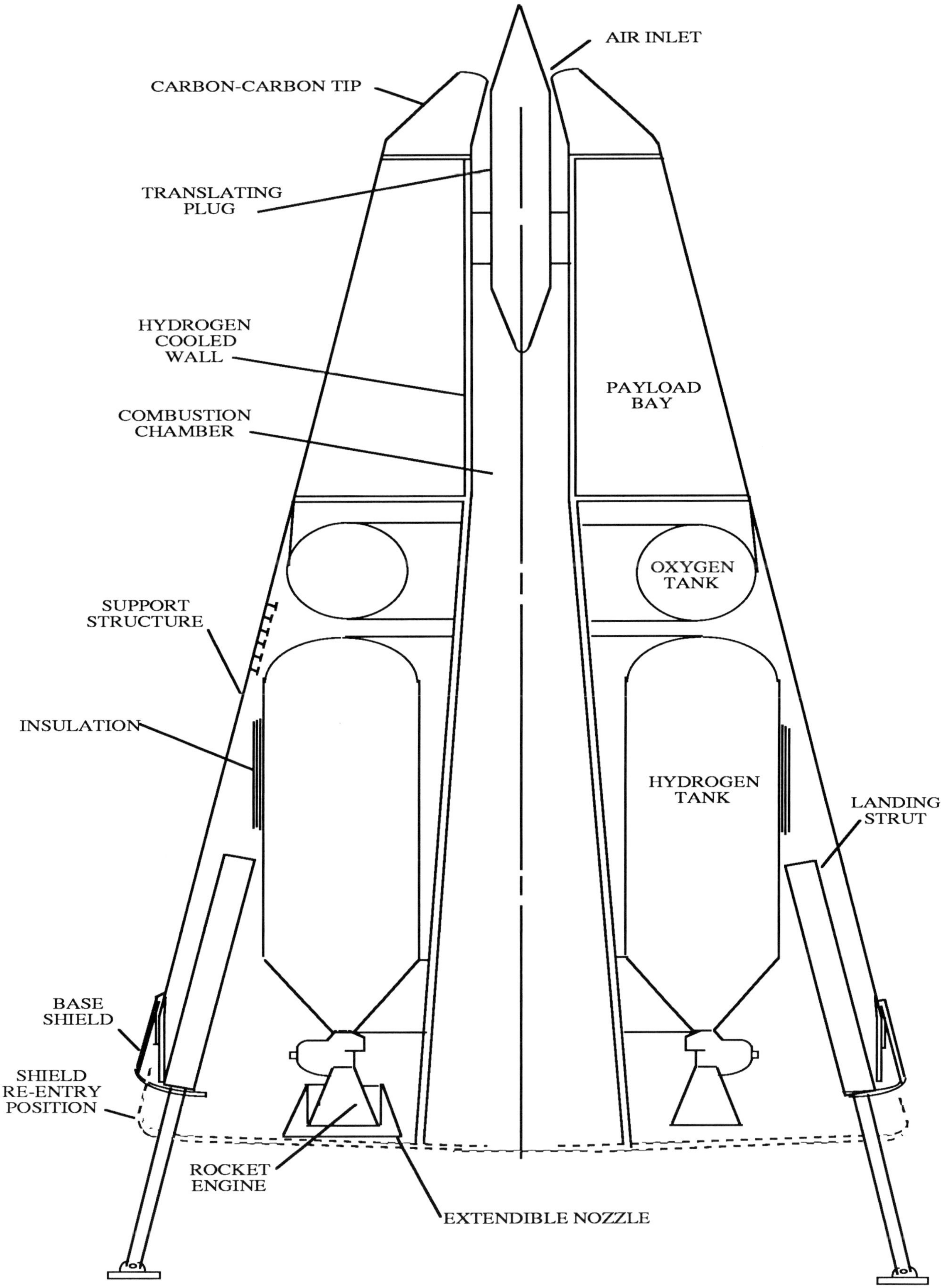

Figure 1 - Conceptual sketch of a scramjet assisted single-stage-to-orbit (Perigee West)

8-5 Outlook

The Delta Clipper development was originally initiated by the Strategic Defense Initiative Office to provide a means of carrying large numbers of small satellites into orbit. The development was authorized in a part of the defense establishment other than the Air Force Space Division that would normally inititiate an effort of this nature. SDIO's plan was to eventually pass on the procurement responsibility to the established procurement authority.

The SDI office was ordered closed in 1993 by the Secretary of Defense. Plans for orbital devices like "brilliant eyes" and "brilliant pebbles" have been largely shelved in favor of ground based ballistic missile defense. SDIO activities, including the SSTO, were reassigned to a new operation, the Ballistic Missile Defense Office (BMDO).

In 1993, the Congress directed that development of advanced space launch vehicles be taken over by ARPA.This was an unexpected turn of events, since all previous development of launch vehicles has been accomplished by the agencies that required them. At that time it was believed that future work on SSTO vehicles will be managed by ARPA. Later decisions, however, put the future of reusable launch vehicles, including the Delta Clipper, in NASA's hands.

The SSTO has now had four different sponsors, of varying committment to the idea. The program has support in the Congress, however. Congressional appropriations for 1994 may result in further evaluations of SSTO's as the future space transportation system, rather than continuing with the Delta Clipper concept. A new launch policy, issued by President Bill Clinton in August of 1994, calls for NASA to begin technology development that could lead to a reusable launch system after the year 2000. The policy also abandons plans for development of a new expendable launch system, long sought by the Air Force, in favor of evolving existing launch systems to greater capability.

9

Safe Rockets - The Hybrid Design

9-1 The Hybrid Idea

During the late 1950's and early 1960's three types of rocket propulsion were investigated in studies that were directed toward determining what systems should be developed for military and civilian applications. Rockets fueled by liquid propellants were already in development for intermediate and long-range ballistic missile programs by the Department of Defense. Other options were confined primarily to solid propellant rocket motors and hybrid motors.

In solid propellant motors, the propellant consists of a mixture of oxidizer and fuel. The propellant is formulated in a pourable mixture and is cast in a "grain" inside the combustion chamber where it hardens to a texture similar to hard rubber. The burn rate of the propellant is determined by the exposed area in the grain that is exposed to burning. Solid rocket motors are hazardous objects and require extreme care during fabrication, transportation and handling.

Hybrid rockets consist of a combustion chamber similar to solid rocket motors, but in this case the propellant is a solid fuel grain. The oxidizer is a liquid and is injected into the chamber at a rate that promotes continuous burning. The combustion rate is a function of the exposed grain area and the rate of oxidizer injection. Unlike a solid rocket motor, a hybrid motor can be shut down and restarted.

Hybrid motors are safe to manufacture, transport and handle. Because the fuel is relatively inert - no more hazardous than wood - a hybrid rocket is unlikely to explode and burn in event of a launch or flight mishap. In this respect, hybrids hold a marked operational advantage over solid or liquid-fueled rockets.

In the course of events, as the studies were completed, the Department of Defense elected to use solid rocket motors for propulsion for its next generation submarine and land-based intermediate and long-range missiles. The simplicity of all-solid rockets for military applications outweighed any safety and performance advantages that hybrid rockets had to offer. Polaris, Poseiden, Trident, Minuteman and Peacekeeper were all designed as multistage solid rockets, as were several smaller Army field rockets.

Hybrid rocketry was therefore shelved for many years, only occasionally appearing in advanced systems studies as an alternative propulsion method. But proponents have remained loyal to the idea, and in recent years research and development work has begun in earnest.

9-2 How Hybrids work

Hybrid rocket motors offer significant advantages over conventional solid-propellant and liquid-propellant rockets. Low development cost, inherent safety, ability to throttle thrust, ability to restart, and environmentally benign exhaust products accrue to place this technology at the forefront for future space launch vehicles. In many respects, hybrids can be considered to be the evolutionary end product of solids and liquids.

A typical hybrid motor uses a solid hydrocarbon fuel cast inside a combustion chamber and a liquid oxidizer, which is stored in a separate tank. The hybrid rocket motor burns like a wood fire, i.e. directly above the surface of the fuel. In fact, a rocket of this type could be a wood burner, or a coal burner, but its performance would not be acceptable. The oxidizer is sprayed into the forward end of of the combustion chamber and is heated initially by an ignition fluid. Hot oxygen vapors flow down the combustion port, vaporizing the fuel which mixes and burns in a combustion flame zone above the fuel. Heat from this flame vaporizes more fuel and thus the process is self-sustaining as long as oxidizer continues to flow. The process is controlled by the flow of oxidizer down the combustion port. With less flow, burning is reduced, thus reducing thrust.

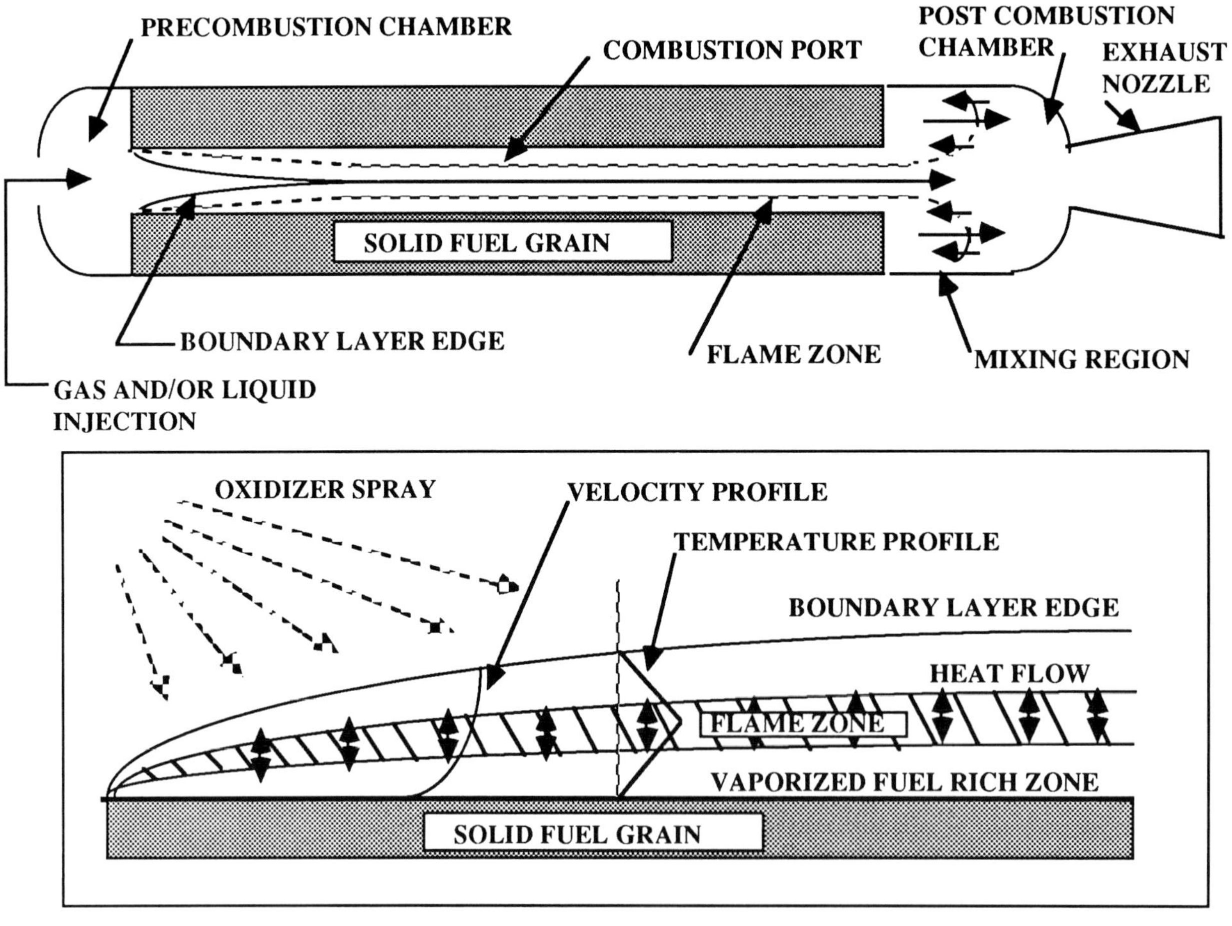

The Fuel Burning Process in a Hybrid Rocket Motor (American Rocket Company)

Hybrid rocket motors possess a unique feature that sets them apart from all other rocket motors. Hybrid motors cannot explode. A typical hybrid motor will utilize a solid synthetic rubber for fuel and liquid oxygen as the oxidizer. These propellants are separated both physically (stored in separate vessels) and by phase, one liquid and one solid. Thus the potential for intimate mixing of the fuel and oxidizer leading to a catastrophic explosion is eliminated. Hybrid rocket motors are not classified as explosive devices.

Safety precautions at the launch sites of conventional launch vehicles are governed by the potential for damage, should an on-site explosion occur. The TNT radius for any launch vehicle can be calculated based on the explosive force that would be encountered if the propellants were intimately mixed together. Hybrid rockets have a TNT radius of zero.

Hybrid rocket motors do not require large expenditures for development. This is in part due to its safe handling properties. But, also, development testing requires fewer test articles. The ability to verify fuel port geometry is unique to the hybrid. Unlike a solid rocket motor, a hybrid test motor can be started and stopped several times, permitting designers to make detailed evaluations at selected points in the burn period. Hybrid motors can be characterized more accurately by this process, which means that development testing can be concluded more rapidly and more economically.

The production cost of a hybrid motor will be significantly less than that of equally-sized liquid or solid rocket due to the hybrid's greater tolerance to manufacturing defects and to its inherent safety. Hybrid rocket motors can be manufactured in a light industry environment with no special safety requirements. To ease shipping loads, it is conceivable that in an operational situation the motors would be loaded with propellant near the launch facility.

Unlike existing solid rocket motors, hybrid motor exhausts contain no hydrochloric acid or aluminum oxide. Local and high altitude environmental effects are on a par with, or may be better than, liquid fueled rockets. The affect of rocketry on ozone depletion is an on-going question. Regardless, the trend for future rocket designs will be to build clean burners.

Hybrid rocket motors have better properties when scaled upward in size. The larger a conventional liquid or solid-fueled rocket is, the greater the explosive hazard, and the greater the cost of development, testing, production and operations. Since hybrid fuel is inert, safety is not a factor in scaling a hybrid motor. A five million pound thrust motor still has a TNT equivalent of nearly zero.

9-3 AMROC'S Venture

The American Rocket Company (AMROC) of Ventura, California, is the world's leading developer and manufacturer of Hybrid Rocket Motors. This entrepreneurial company is dedicated to development of hybrid rocket motors for space launch vehicles. AMROC is a privately funded company, a testament to the fact that rocketry is coming of age as a mature industry and no longer needs to seek development under government sponsorship.

AMROC has tested several classes of hybrid motors with thrust ranges ranging from 60 pounds to 260,000 pounds. The programs range from high-thrust booster motors to small storable propellant upper stage motors. Extensive development has taken place on the H-500 (75,000 pound thrust) hybrid propulsion system.

Major components of AMROC's hybrid motors, which use a solid fuel and a liquid oxidizer, are the combustion chamber, lined with fuel; an exhaust nozzle fitted with lateral injectors for steering; an oxygen injector with main propellant valve; and an oxygen tank. The oxidizer may be pressure-fed or pump-fed to the combustion chamber. Pump-fed systems are lighter in weight than pressure-fed, but may be more costly. A pump requires a gas-driven turbine to drive it, along with its own propellant supply.

1993 test firing of AMROC's 250,000 lb thrust hybrid rocket motor (American Rocket Company)

AMROC's most recent development is the H-1800, 260,000 pound thrust hybrid motor, which was successfully tested at Phillips Laboratory, Edwards Air Force Base, California, in January, 1993. The H-1800 motor development program began in 1991 with successful firings of the H-100 propulsion system in a "two-port" system. These tests verified the combustion characteristics of the full-scale H-1800 motor.

This testing technique is an important aspect of the promise of low cost in the development phase of a hybrid rocket. Because the propellant grain is symmetrical, it is possible to perform realistic scaling tests on as few as two ports of the full scale chamber and obtain credible results for scaling from one thrust level to another.

AMROC's H-1800 motor is the largest hybrid propulsion system being developed today, delivering nearly 19 million pound-seconds of impulse. The vehicle uses liquid oxygen for its oxidizer and solid polybutadiene (HTPB) for fuel. The oxygen tank is pressurized by helium gas to deliver oxygen under high pressure to the combustion chamber. The H-1800 is a robust, safe propulsion system, demonstrated to be economical by reason of its short development time. While cracks in the propellant grain can be disastrous for conventional solid motors, in a hybrid motor a crack cannot cause accelerated burning, because only fuel rich gasses flow over it.

The H-1800 motor case is a filament-wound graphite/epoxy composite case. Polybutadiene fuel is cast into an internal grain configuration that resembles a wagon wheel, where the fuel elements form the spokes of the wheel.

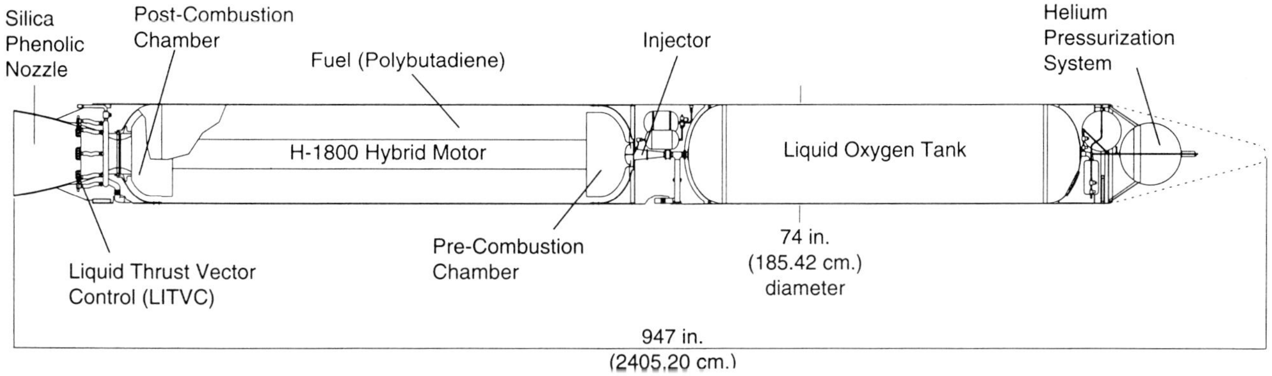

AMROC H-1800 Hybrid Rocket Motor (American Rocket Company)

9-4 AMROC's Aquila Rocket

American Rocket Company's bid for commercial launch business centers on their Aquila Concept. The Aquila system is directed toward meeting the needs of small satellite builders during the 1990's and beyond. The company believes that its concept will be cost competitive as well as offering the benefits associated with safety, environmental compatibility, and routine operations.

Aquila is based on AMROC's new rocket motor, the 260,000 pound thrust H-1800. It will be capable of placing a maximum of 2000 pounds into polar orbit at 100 nautical miles altitude. Launches will initially be from Vandenberg Air Force Base, California. Current planning calls for the first flight in 1996.

Aquila 21 employs two H-1800 liquid oxygen/HTPB hybrid propulsion systems for the first stage, one H-1800 for the second stage, one United Technologies Orbus 21S solid rocket motor for the third stage, and one U-75 pressure-fed nitrous oxide/HTPB hybrid propulsion system for the fourth stage. Use of the storable propellant nitrous oxide for upper stages is an indication that AMROC has not arrived at acceptable methods of storing liquid oxygen, a cryogenic propellant, for moderately long periods in its upper stages, opting instead to use an oxidizer that stores well at room temperature. A mix of different propellants for its hybrid stages, plus use of a solid propellant rocket motor for one of its stages, tends to weaken AMROC claims of safety, reliability, and economy that might otherwise be true for an all hybrid system.

AMROC has three major vehicle developments under way, the Aquila 21 small launch vehicle, the Aquila 31 and the Hy Flyer Launch Vehicle. The Aquila 31 is similar to Aquila 21 with the exception that it uses three booster rockets instead of two. Its payload is 2650 pounds to low-earth polar orbit.

The HY Flyer is a non-orbiting launch vehicle designed to carry payloads weighing up to 7000 pounds to high altitude. Each flight will provide over ten minutes of micro-gravity time. Hy Flyer uses a single H-1800 hybrid motor topped by a nose cone containing a payload module and a payload recovery system. The first flight will be in 1994, carrying a NASA payload.

Various studies of heavy-lift launch vehicles sponsored by both NASA and the Department of Defense have included core vehicles with strap-on booster rockets. Candidate strap-ons have been solid rockets, liquid-fueled rockets, and hybrid rockets. AMROC plans to stay abreast of developments in this area with the intent of pursuing hybrids as the preferred candidate in this application.

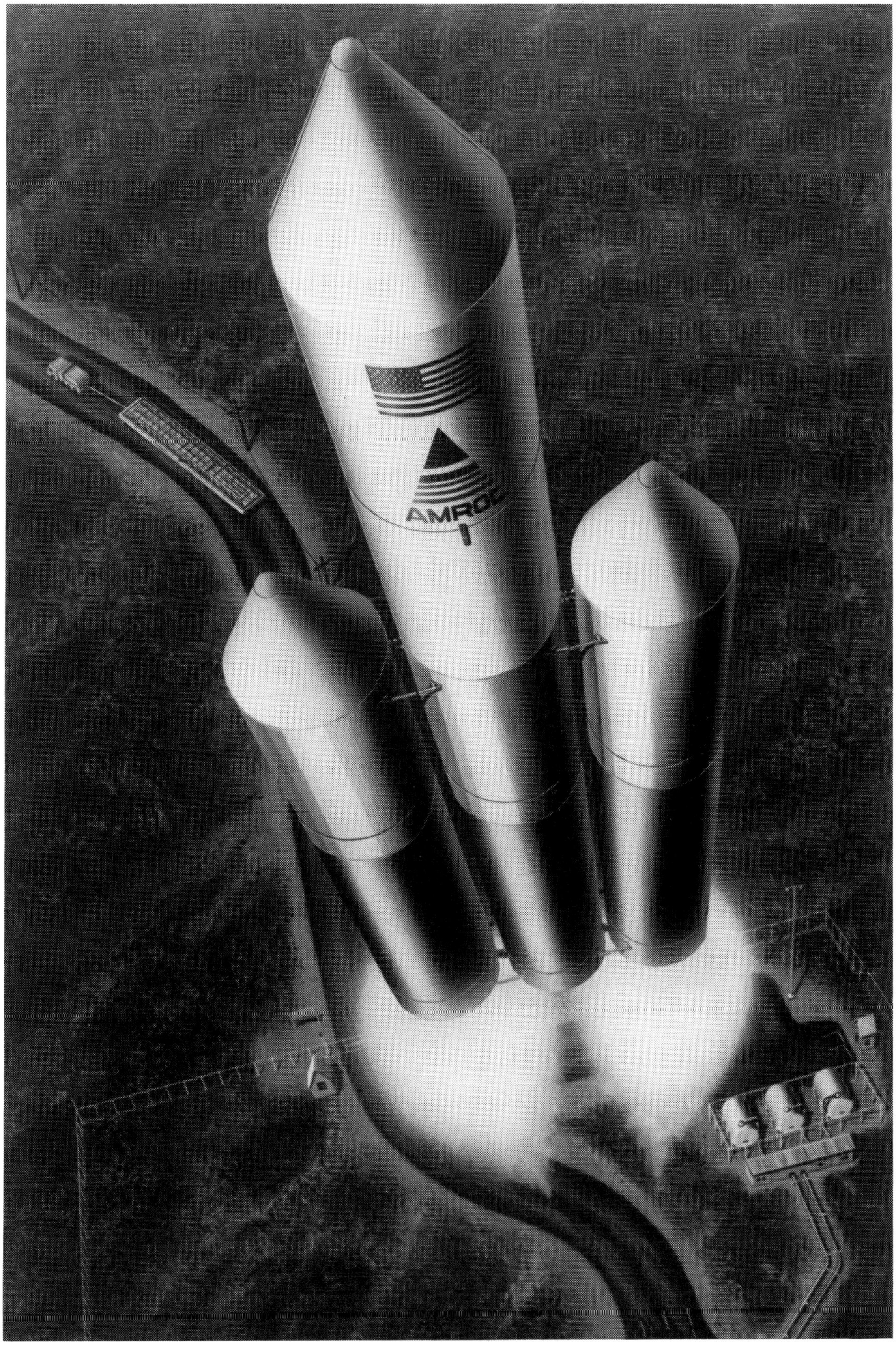

Artist's concept of the Aquila 21 Hybrid Launch Vehicle lifting off at Vandenberg Air Force Base (American Rocket Company)

9-5 The Sea Based Launch System

A synergy of sorts exists between hybrid rockets as a safe launch system and a sea-going vessel as a launch platform. A hybrid rocket could be launched from a ship without fear of catastrophic failures. The ship in turn provides advantages to launch operations that cannot be matched by land-based operations.

A launch ship can launch from virtually any point at sea. It is an all-azimuth launch platform. A ship has the advantage of being able to move away from weather, thus is better able to meet launch schedules. It is free of the launch priorities that are common to based complexes. It is free of concerns regarding how a failed launch could affect the local population. At Vandenberg Air Force Base, Titans can be launched only when wind conditions are right. (If the wind is not right, a catastrophic failure could result in blowing clouds of toxic gases over populated areas.) A ship can travel to the equator to make the most efficient launches to geosynchronous orbit.

In the future, when space programs employ nuclear power and propulsion systems, it will be a constant battle with environmental interests to obtain approval for launching from land bases. A sea-based launch system will alleviate these concerns.

In 1989 Perigee West Company of La Jolla, California, performed a systems study for the Air Force that analyzed a hybrid launch system integrated with a large ship.A set of ground rules was established to bound the scope within the available funding.

The geographic location was confined to the west coast with basing in San Diego, California. Baseline performance was set at 25,000 pounds of payload to low-earth orbit in a polar inclination. Only hybrid launch vehicles were considered, including upper stages, due to the safety issue. Flight rates were assumed to be one every two weeks, and geosynchronous missions were to be launched from the equator.

It was recognized at the outset that feasibility of the idea could only be ascertained if the idea were examined on a total systems basis. Ship design, launch systems and support systems were therefore given equal emphasis during the study, with the result that a fully integrated system concept was produced.

The study results showed that hybrid rockets in combination with sea-based operations can provide a dramatic improvement in space launch capability, showing lower development and recurring cost, unmatched flexibility, rapid response to requirements, and marked improvement in safety of operations.

The hybrid launch vehicle configuration concept developed for the Sea Based Launch System (SBLS) is a multistage arrangement using same-diameter fuel chambers in all stages. The boost stage comprises a cluster of five fuel chambers, each chamber rated at a thrust level of 465,000 pounds. The cluster is separated at boost burnout.

The second stage comprises the four outer fuel chambers of a cluster of five, each chamber rated at a thrust level of 176,000 pounds. The cluster of four chambers is jettisoned along slip rails at the conclusion of second stage burn.

The third stage consists of the central thrust chamber at the same thrust level, coupled to an equipment module at its forward end, which in turn is connected to the oxygen tank. A single, lightweight oxygen tank contains oxygen for all three stages of the rocket. The payload fairing is attached to the forward end of the oxygen tank.

The equipment module contains turbopumps for feeding propellants to the fuel chambers, and avionics for control of the vehicle.

The SBLS hybrid rocket is, in fact, a "quasi 2-stage" vehicle, the boost and second stages constituting half stages and the third stage consisting of a full stage.

Perigee West's Sea Based Launch System during launch of a hybrid rocket

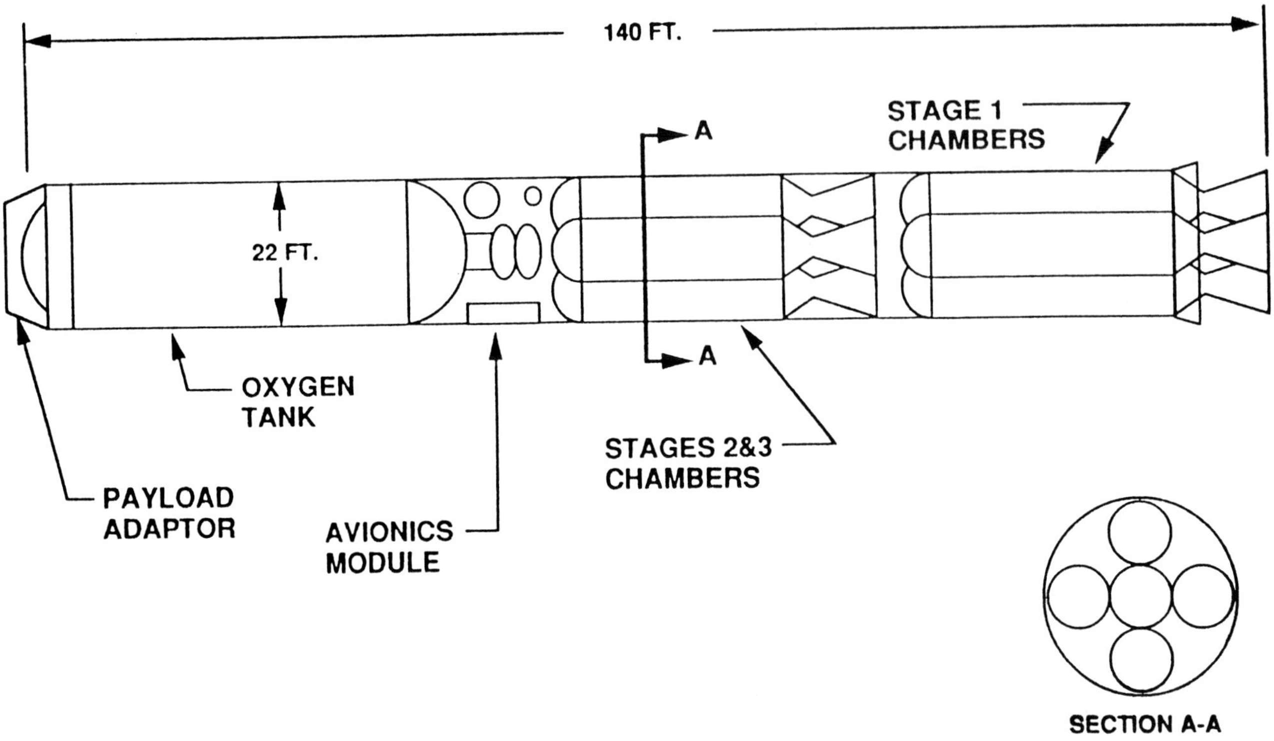

The baseline hybrid launch vehicle conceived by Perigee West for the SBLS

An attractive feature of hybrids is the ability to verify burn in all chambers before liftoff by operating for a short time in a low thrust or idle mode. Unlike conventional solid motors, hybrids can be shut down if they are not all operating properly. Once operating, because of its tolerance to defects, there is little reason to expect a solid fuel chamber to shut down during flight. Historically, curtailed flights of solid propellant rockets have been accompanied by catastrophic failure, mostly due to defects of one type or another in the solid grain, or in its bond to the outer shell.

The individual hybrid motors used in the SBLS are seven feet in diameter. The starting device is a cartridge containing a pyrophoric fluid, which ignites spontaneously with oxygen. The starter is located on the central oxygen injector and the flame is propagated radially to the remaining fuel ports.

Steering of the hybrid vehicle is achieved by injecting liquid oxygen into the exhaust nozzles. Eighteen electrically actuated valves surround each nozzle on an equal spacing. Technology for this method of steering is drawn from the Titan vehicles, which use a similar system.

For missions to high orbits, SBLS utilizes an additional transfer stage. It too is a hybrid vehicle, designed for high altitude operation. It is similar in principle to the lower stages with the exceptions that it is a pressure-fed system and steering is accomplished by gimbaling the exhaust nozzle.

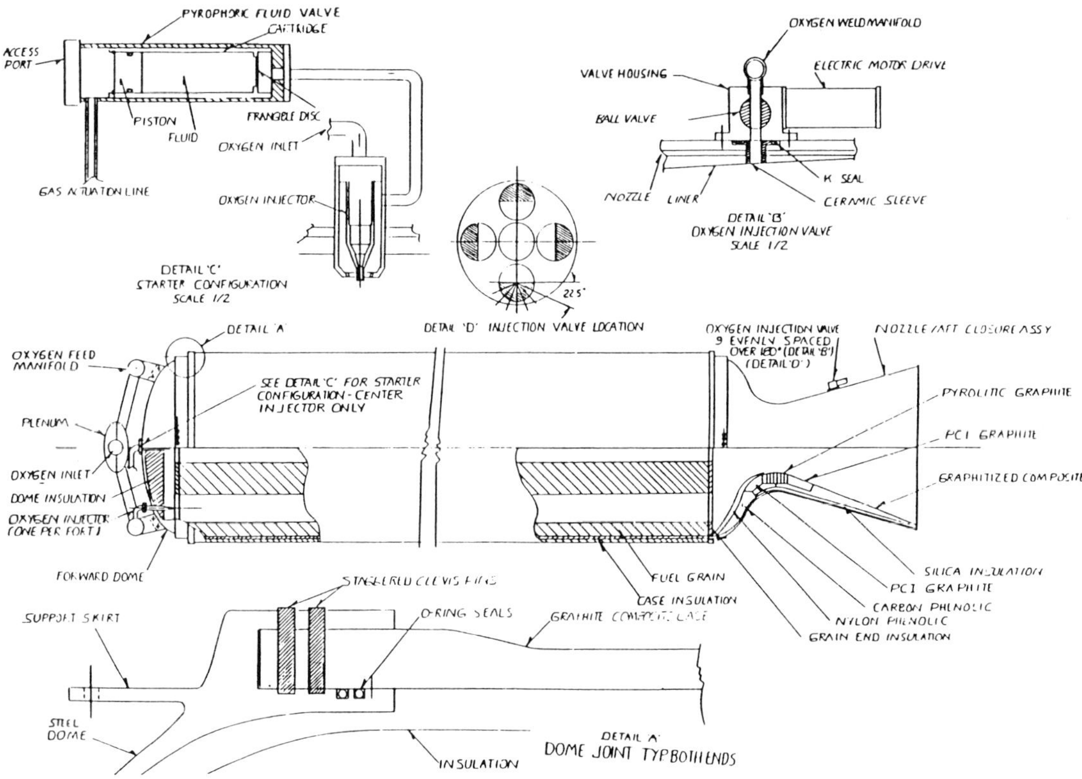

The SBLS Hybrid Rocket Motor

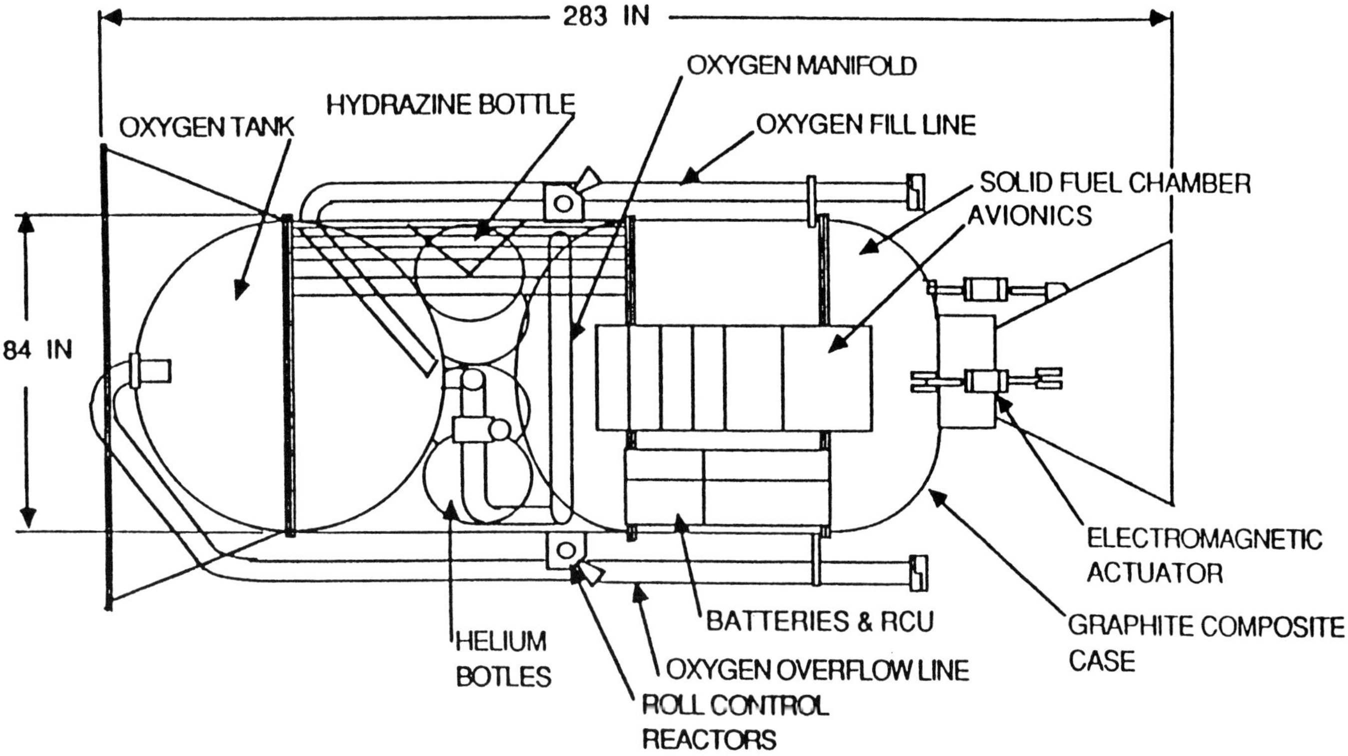

A Schematic Diagram of the SBLS Transfer Stage

To be effective, the SBLS requires a launch-ship profile that affords the maximum possible area and space for operations, requires minimum maintenance, can be operated by a small crew, and has reasonable speed for transit between home base and launch operations.

A ship design similar to those used for carrying crude oil meets these criteria, modified to accomplish the desired vehicle assembly, checkout and launch services. Tanker ships may measure 240 feet in beam by fourteen hundred feet in length. These ultra-large carriers were originally designed to carry oil around the Cape when the Suez Canal was closed.

A ship about the size of the Exxon Valdez, 170 feet in beam by a thousand feet in length, was selected for the SBLS due to this size being the upper limit of fabrication capability for most U.S. shipyards. The tanker design was altered to provide a large over-deck superstructure where most of the launch vehicle final assembly and checkout takes place.

Special provisions in the ship duplicate the capabilities of ground facilities that service launch vehicles at the East and West coast land-based launch sites. These include storage of liquid oxygen and pressurant gases required to service the hybrid rockets, a launch control center, a payload processing facility, laboratory and shop support, and a launch platform. The SBLS is essentially "Cape Canaveral on a Ship."

Dynamic analyses showed that the sea-based system is sufficiently stable for launching over the most prevalent sea states. Ability to move away from the weather adds a significant element of flexibility to this method of launching payloads.

An essential element of the system is a port facility designed to accommodate the servicing needs of the vessel, as well as provisioning the partially assembled launch vehicles. The 24th street terminal in San Diego harbor showed itself to be adequate to this purpose, with availability of large empty buildings and adjacent open-land area. In the SBLS plan, components of the launch vehicle system would be shipped to the terminal where they would be assembled and held ready for transfer to the ship. The ship itself is capable of holding four vehicles in a launch readiness condition. Polar launches based on the San Diego system could be initiated within forty-eight hours of demand.

Hybrid launch vehicles are, of course, not confined to sea basing. The study simply explored this option and disclosed the many advantages of such a system.. Land based hybrids can compete effectively with conventional solid propellant and liquid propellant rockets. Promise of lower cost of production and the added advantage of safety will eventually propel this mode of rocketry to the forefront of U.S. launch systems.

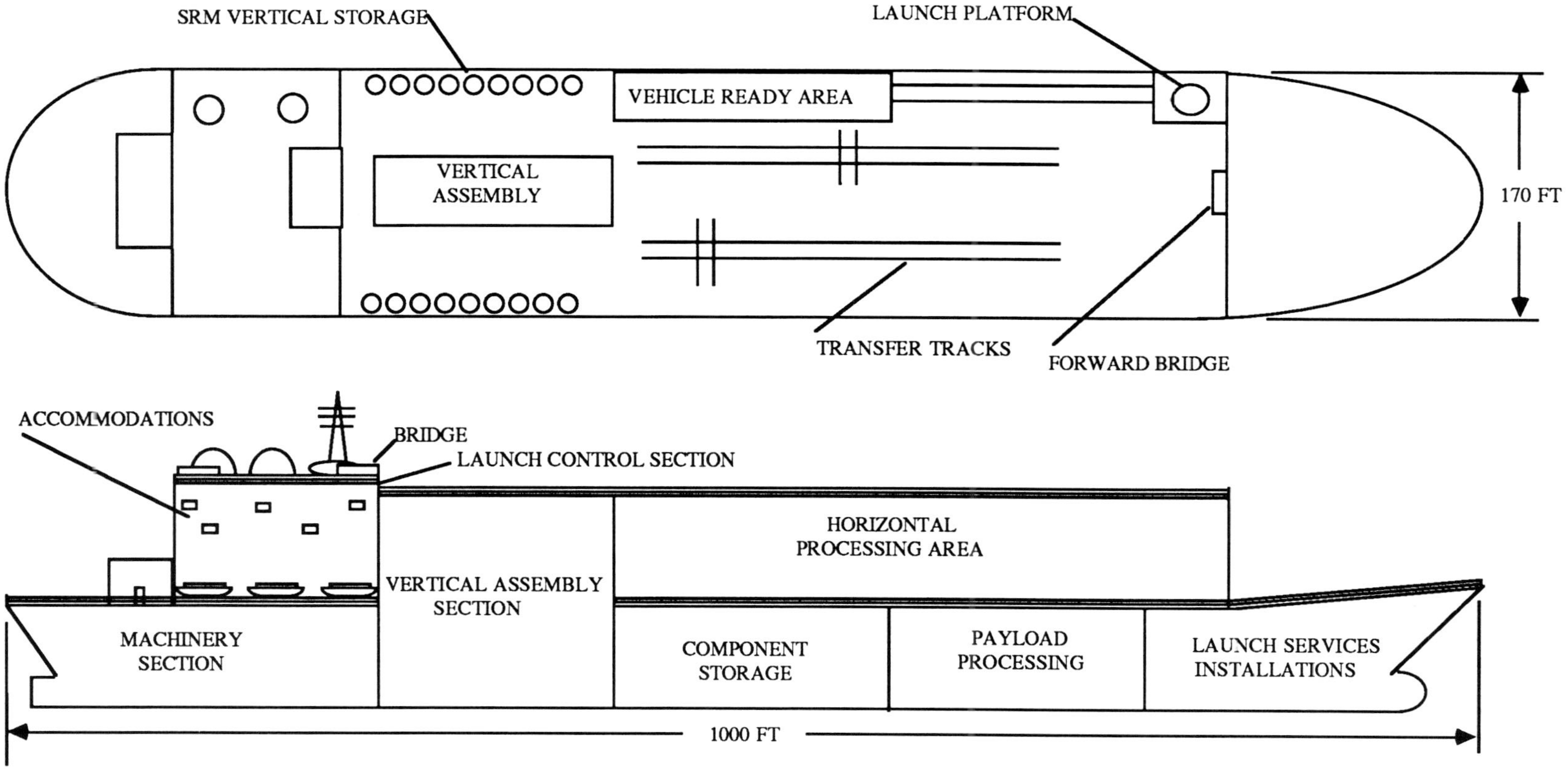

Side and top profiles of the SBLS launch ship (Perigee West Company)

9-6 Other Hybrid Developments

At NASA's Marshall Space Flight Center, where most of NASA's propulsion activities are centered, there is a growing interest in hybrid rocket development, as evidenced by several commitments in the past few years to begin development work. Most were set aside because of funding shortages, but new efforts are beginning to take hold.

NASA's pursuit of hybrids will characteristically be more conservative than that underway at the American Rocket Company, where significant progress has already been made in developing large rocket motors. Whereas AMROC's aim is to penetrate the commercial market for space launches at the earliest possible date, NASA's concerns will initially be to achieve a detailed characterization of all the elements required to understand hybrid technology. It is interesting that the industry view with respect to hybrids is that rocket design is a mature technology and can now be pursued independently of the government, whereas NASA's view is: "Hold on, we have to understand all this first before proceeding into hardware construction."

To this end, future studies will be directed toward selection and characterization of propellant formulations, development scaling laws, understanding of fuel-grain geometry and regression characteristics, oxidizer injection and ignition systems design, combustion kinetics, combustion chamber insulation formulations, thrust-vector control methods, and testing with flawed materials. In late 1991 and industry-government team consisting of Marshall's Propulsion Laboratory and industry members Thiokol, AMROC, General Dynamics, Martin Marietta, Rocketdyne, and United Technologies was formed to advance hybrid rocket technology. Testing of small scale oxygen/polybutadiene motors started in May, 1992.

At JPL, researchers have been conducting numerous tests with a slab burner motor to assess burning rate of some fuels under different conditions. Slab motors use flat slabs of fuel in a motor with gaseous oxygen introduced at the head end. Burn rate is determined by carefully weighing the slab before and after a burn. JPL has tested a variety of fuels, including some that contain aluminum and coal.

Hybrid rocketry is also being studied at several universities as part of their research projects and teaching curricula. Students at the Air Force Academy designed, built, and launched hybrid rockets as part of a rocket propulsion course. MIT, Purdue, the University of Arizona, the University of Alabama at Huntsville, and a consortium of Utah universities all have programs underway to test hybrid motors.

9-7 Outlook

It was stated at the beginning of this chapter that hybrid rockets may be the evolutionary end product of liquid-propellant and solid-propellant rocket systems. Given the present lax state of rocket launch vehicle development, insofar as congressional support and available funding are concerned, this conclusion may be slow to be realized

Rapid progress and spectacular success on the part of AMROC in their private venture to secure a share a of the commercial launch market could accelerate adoption of hybrids by NASA and the Department of Defense.

The Air Force plans to fund another round of industry studies to define the next generation of launch vehicles. Having previously failed in a joint effort with NASA to obtain approval to develop a "National Launch System," the agency has revised its objectives and is starting again with clean-sheet approaches. Should hybrids appear as a strong contender in these studies, they could star in follow-on developments.

A hitch in the plan has come about as a result of recent Senate and House appropriation committee determination that any new launch vehicle effort will be undertaken by ARPA. As a result, funding for studies in this area were allocated to ARPA instead of the Air Force in Novemnber of 1992. The ultimate result, however, will likely be that ARPA will enlist Air Force assistance in the effort, as ARPA is not staffed to undertake tasks of this magnitude.

Two long range justifications exist for hybrids to play a strong role future launch vehicle definition. Development of a small personnel carrier to replace the Space Shuttle will require a new lift vehicle. If a safe rocket is possible, then it behooves the government and concerned agencies to move developments in that direction. Hybrids are the only candidate for a safe rocket system.

Secondly, in the future, when nuclear power plants and parts of nuclear propulsion systems are orbited, there will be a strong resistance to launching from land bases like Vandenberg Air Force Base and Cape Canaveral. The hybrid rocket, in combination with sea launching, as described in the SBLS concept, will make these types of launches acceptable to the public.

As a result of the AMROC developments, hybrids are one of only two or three efforts underway in the U.S. directed toward advanced rocketry. In actual hardware development, hybrids are probably ahead of the pack.

10

A Modern Space Launch System

10-1 The Beginnings

In the mid-1980's, The Defense Department and NASA joined forces to determine the nation's space transportation needs in the coming years. The agencies were interested in setting the foundations for the nation's next space launch system. Aerospace contractors were funded to conduct what was termed the "Space Transportation Architecture Study (STAS)."

Justification for the study was the high cost, poor reliability, and poor operability of the current fleet of launch vehicles, consisting of the Space Shuttle, Titan, Atlas and Delta. These systems were viewed as old technology, despite the fact that they were continuously being modernized.

STAS was a two year study. Heavy emphasis was placed on analysis of mission models, examining methods of operation and estimating life cycle cost at high, medium, and low rates of usage. The study examined launch system options and technologies required to achieve them.

The study was flawed from the beginning by the assumption on the part of both NASA and the Air Force that the new launch system was to center on heavy payload lift capability. The mission models were heavily influenced by prospects of launching significant numbers of heavy payloads for the Strategic Defense Initiative (SDI). At the same time, little attention was given to the payload range in which Atlas and Delta operate, where most payloads lie. Such payloads were assumed to be launched several at a time on the big launchers, a situation that is often logistically impractical for more than two payloads.

But the studies' overall conclusions were that the nation needed a new system to meet future requirements. It would be a robust system, avoiding temptation to extract the maximum performance that materials technology can afford. System operations would be optimized, emphasizing cost reduction in production, processing, and launching the vehicles. The launch vehicles would be built tolerant to failures. The new system would be responsive, ready to fly on short notice as compared to months of processing time for existing launch vehicles.

The study solidified a philosophy. The new system would incorporate high design margins, fault tolerant components and modern methods and processes. An interesting conclusion reached by the contractors, apparently in recognition that future launch systems should be environment friendly, was that the new launch system should be fueled by liquid hydrogen and liquid oxygen.

In July of 1987, following completion of the Architecture Study, the Air Force awarded seven $5 million one-year concept development contracts to industry to expand on the budding ideas for a new launch system. NASA continued to participate in furtherance of their interest in a heavy lifter for future payloads. The new initiative was designated the Advanced Launch System (ALS).

The ALS studies covered all aspects of launch vehicle design, ground support systems design, manufacturing processes, operations, and software. Supporting technologies included new ways to construct propellant tanks, line replaceable avionics, and low cost propulsion. In keeping with the new policy of robust capability, the launcher was required to operate with one engine out, neglecting the fact that a measure of unreliability was thereby introduced through added system complexity. Configuration trade studies identified a requirement for a new engine at a thrust level of 670,000 pounds. The studies produced concepts for a range of launch vehicle sizes that could be readily assembled from common modular components. NASA took on the responsibility for developing the new engine.

The ALS was on shaky ground almost from the beginning, largely due to the difficulty of identifying a lift capability corresponding to an actual requirement. Strategic Defense Initiative requirements disappeared when the large payloads associated with that program ceased to be of interest. The Air Force and NASA failed to find common ground for a system that would meet both their requirements. Shuttle C was being concurrently promoted by NASA as a heavy-lift vehicle that could satisfy their requirements. This activity undermined what might otherwise have been an earnest effort to find a compatible launch system.

Then, too, the ALS did not receive the support of launch vehicle contractors that might have been expected for an new development opportunity. Instead, the ALS was viewed as a threat to established launch systems, all of which would eventually be displaced by the ALS. As a consequence, the Congress was not lobbied enthusiastically by the contractors regarding how the nation would benefit from a modern launch system. The opinions of Congressmen were shaped largely by their own technical staffs.

Congressmen were unimpressed with the ALS when it was presented to them. The high cost of development, which approached $17 billion, was a particular stumbling block. With no pressing need for the ALS on the horizon, funding for the program was reduced to a small amount for a low-level technology effort. The principal surviving effort was the development program NASA had underway at Marshall Space Flight Center for the ALS main engine.

By December of 1990, the Committee on the Future of the U.S. Space Program, chaired by Martin Marietta's CEO Norm Augustine, had issued its report on recommendations for the nation's future space programs. One of its recommendations was to phase out the Space Shuttle by the year 2010 and replace it with a new launch system embodying ideas developed in the ALS studies.

The program acquired a new life, now under the name National Launch System (NLS). At President Bush's direction, the NLS was to be a joint NASA/DoD effort. The widely disparate needs of the two agencies, in terms of vehicle configurations and facilities, had never been reconciled. The result was another program of growing complexity whose costs were rapidly approaching that of the original ALS. A big problem was that the specialized requirements of the two agencies remained somewhat in the dark, rendering it difficult to agree on the common elements of the system.

When funding requirements for the program at $250 million for fiscal year '92 were revealed, a reluctant Congress decided that more information was needed before agreeing to proceed. The price tag for full deployment of the NLS, as described to the Congress, would be a minimum of $8 billion and possibly as high as $12.2 billion. Funding was approved in an amount that allowed continuation of the engine development program at MSFC and operation of the NLS program office. By July of 1992, faced with mounting budget problems, Congress cancelled further work on the NLS. Only the development work on the engine was allowed to continue at a low level of funding at MSFC.

10-3 An Economical Approach to a Modern Launch System

New investigations to define the launch system best able to meet the nation's future requirements will inevitably revisit studies and development work that have been undertaken at various times during a half century of U.S. rocketry. Key questions that must be addressed are whether the system can be economically developed and operated, and whether it will be capable of accomplishing all missions covering the desired payload weight and size range.

Estimates for development of a reusable SSTO range from $5 billion to $20 billion. The lower estimates have been shown to neglect the cost of fleet acquisition. Whether a reusable SSTO will be a cost effective launch system is a troubling question in view of the Space Shuttle experience. That system has greatly exceeded the cost of expendable launch vehicles on a cost per launch comparison, whereas it was originally expected to displace expendable launch vehicles by reason of predicted lower cost. The challenge for proponents of a new reusable SSTO is not only to close the cost gap between the Space Shuttle and expendable launch vehicles, but to do considerably better.

Proponents of a two stage reusable launch system are the skeptics regarding feasibility of a SSTO. The two stage approach faces the same cost challenge as the SSTO, but under an added handicap, since the system consists of two complete vehicles instead of one. The system concepts under consideration involve winged vehicles for both booster and orbiter. Development cost could exceed the cost of developing an SSTO.

Upgrading the existing fleet of launch vehicles has strong support as a means of holding down costs in an era of tight budgets, and appeared in early 1994 to be the top recommendation stemming from Air Force and industry investigations during three months of study in Colorado Springs. With the recent Martin Marietta purchase of General Dynamics Space Division, the scenario for this approach could narrow down to improving the Delta and Titan systems. As Martin will be the owner of two vehicles with similar capabilities, the Titan core and Atlas, economics would appear to favor one or the other of these vehicles. Centaur will be the upper stage in either case. Delta already has a flight reliability approaching 100%, so it would appear that upgrade expenditures would be directed to the Titan vehicles, whose success rate for the last 100 launches is about 94%. Acquisition and operations costs are not likely to decrease as a result of the upgrade effort. In fact, costs will probably increase due to loss of commercial flights to foreign launchers.

An expendable launch vehicle based on advanced technology can follow one of two approaches, SSTO or multi-stage. The multi-stage is generally favored because it better addresses the problem of efficient handling of a range of payload weights, using small versions for small payloads and larger versions for heavier payloads. An SSTO would have to be built oversized to manage the full payload weight range of interest. The main drawback to a high technology launch system is development and acquisition cost. Also, operations costs lower than those currently being experienced cannot be predicted with confidence.

A final option, developed by Perigee West Company of La Jolla, California, addresses a method of acquiring a modern launch system with low risk and at low cost by identifying the parts of launch systems that have worked best in the past and leveraging the design of the new system on those selected features and components. The new system would be produced under production and operatiions disciplines that incorporate advanced methods and processes already demonstrated to have been effective for lowering cost and improving quality.

In its search for a method of launching payloads at reduced cost, the government was ill-served by industrial contractors, who omitted investigations and trade studies that would have identified the right kind of launch vehicle during the ALS and NLS studies. Instead, investigations concentrated on how to build the cheapest hydrogen fueled vehicle, which side-stepped a principal issue. Had the trade studies been done, a hydrogen fueled system would have fallen by the wayside. It would have been displaced by a system fueled with RP-1, a fuel similar to kerosene.

There are a number of reasons why an LOX/RP-1 launch system will be more economical to buy and operate than a LOX/hydrogen system. To begin with, the launch vehicle is much smaller, containing a fuel that is thirteen times more dense than liquid hydrogen. Smaller size means less material to manufacture and easier shipping and handling, all of which contribute to lowering overall cost. A hydrogen fueled launch vehicle is considerably more complex than a RP-1 fueled vehicle. Tanks and propellant lines require special insulation. Operating procedures must be contrived to eliminate possibility of moisture injestion. Special chilldown operations are required to condition the turbopumps before startup. Ground operations and equipment have to support processing of a deep cryogenic propellant, as opposed to a mild, ambient temperature fuel. When all is done, comparative predicted performances of hydrogen fueled and RP-1 fueled launch vehicles have always fallen within 10 % of each other, sometimes showing hydrogen to be better, and sometimes favoring RP-1.

RP-1 fueled rocket engines are inherently simpler than hydrogen fueled engines and therefore should be lower cost. At more than $38 million each, the Space Shuttle main engine illustrates clearly the trouble the government can get into by selecting this type of engine for an expendable launch vehicle. Admittedly, a problem with rocket engine production that contributes to high cost is production in what is essentially a laboratory environment. On the other side of the coin, Rocketdyne's production of Atlas MA-3 LOX/RP-1 engines in the early 1960's at its plant in Neosho, Missouri, amply demonstrated that such engines can be manufactured economically in a semi-automated production shop.

A LOX/RP-1 launch system is a relatively easy development. By leveraging heavily on equipment designs that have demonstrated very high success rates, much of the risk associated with stepping into an all-new design is eliminated. This yields direct benefits in shortened development time, lower cost, and lower cost of development.

As part of leveraging on past design experience it is beneficial to examine the success history of the three major launch systems now in operation; Delta, Atlas and Titan. Examination of flight histories reveals that the highest success rate has been achieved by the Delta launch system, which currently enjoys a string of over fifty successful launches. Examination of the success rate of the Delta and Titan *core vehicles* shows a success rate close to 99%. One can conclude then, that the technology embodied in the Delta and Titan core stages is proven and represents the best possible takeoff point for the design of a new launch vehicle. Toxicity and high cost will rule out Titan propellants. The structures in both systems are robust, however, and are a proper guide for the design of a new vehicle. A prudent and practical approach would take these two designs and leverage them through improvements in methods and processes into a design that will be suitable for the proposed modern launch system.

For propulsion, the RS-27 engine on the Delta has had nearly a 100% success rate. It makes good sense to leverage this design into an engine that would be suitable for the new launch vehicle. Rocketdyne has produced a concept for doing this in its RS-X engine, which would be a relatively easy development. While fully adequate to the new launch vehicle's initial requirements, the concept invites a logical improvement path that eventually leads to significant vehicle performance gains, should it be necessary in the out-years.

The core vehicle for the proposed new launch vehicle is fueled with liquid oxygen and RP-1. It consists of a robust structure leveraged on technology embodied in Delta and Titan core stages and is powered by four Rocketdyne RS-X engines delivering a total liftoff thrust of 1,696,000 pounds. The core stage measures 18 feet in diameter by 103.5 feet in length. The illustration shows a common bulkhead between the two tanks, following the trend of a Martin Marietta effort to improve the efficiency of the Titan core stage. Such a bulkhead would be stiffened to withstand reversal pressures over a limited range.

Advancements in methods and processes to be incorporated in the core vehicle include use of lithium aluminum alloy for lighter weight structure and tanks, automated x-ray inspection of welds, pulse plasma welding, CAD-CAM fabricated adapters, and spin-formed, one-piece bulkheads.

The core stage retains the use of fluid operated engine gimbal actuators, since this is proven, economical technology. Electromechanical actuators were favored for the hydrogen fueled NLS. The alternative for the NLS was to install a separate hydraulic power supply for operation of gimbal actuators, whereas on a RP-1 fueled vehicle high pressure fuel from the turbopumps is available for their operation.

Guidance, control, telemetry and other electronics (avionics) are always carried on the upper stage. An interesting aspect of the avionics section of launch vehicles is that the industry has been forced into a policy of continuous modernization of equipment, so in any given year it can be generally characterized as "state of the art." This is due to the rapidity of change within the electronics industry, a consequence of which is a reluctance by manufacturers to produce obsolete electronics. When required to do so, the cost is very high.

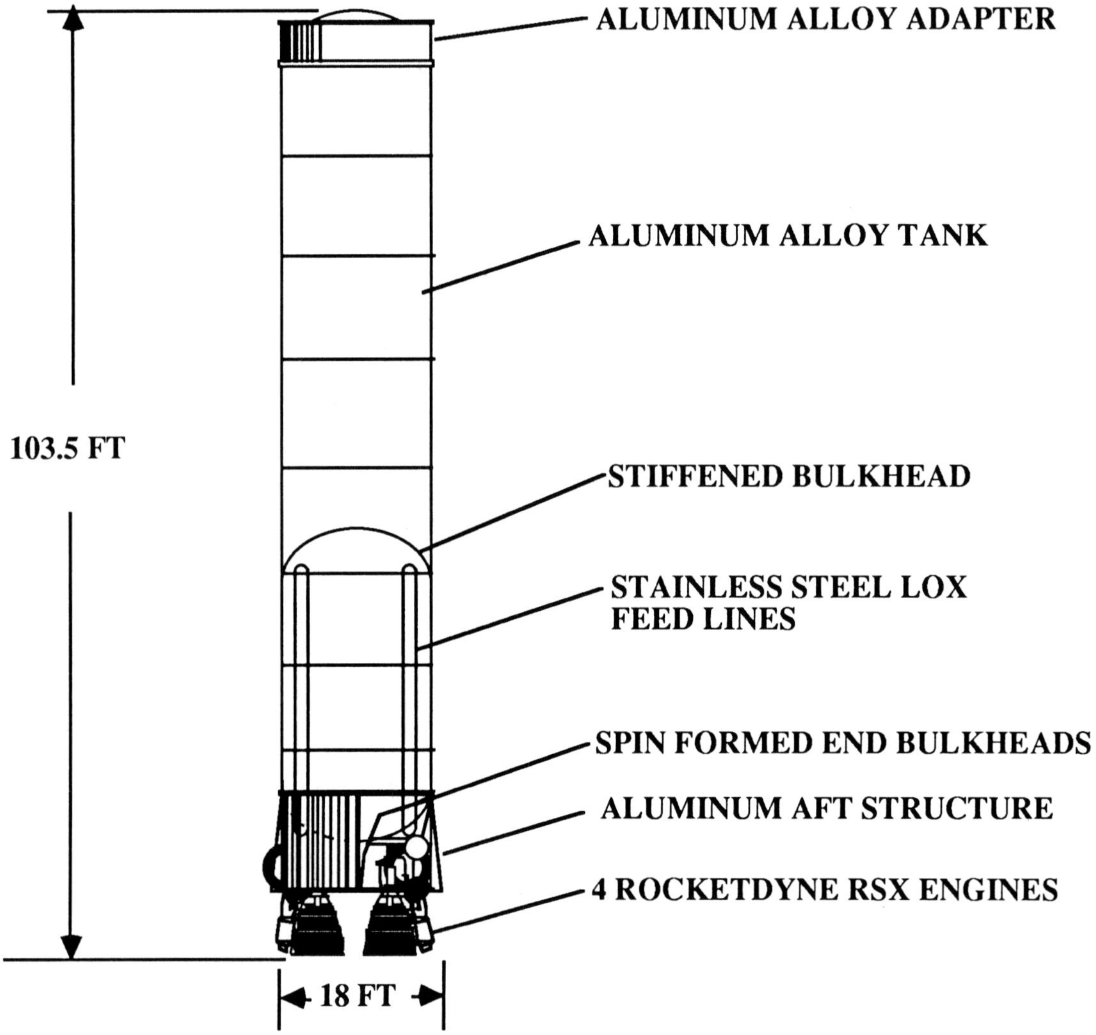

PROPELLANTS - LOX/RP-1

MIXTURE RATIO - 2.25

DRY WT. - 80,000 LB

LIFTOFF THRUST - 1,696,000 LB

LOX TANK PRESSURE - OXYGEN

FUEL TANK PRESSURE - HELIUM

SEA LEVEL ISP - 264

ALTITUDE ISP - 299

Core vehicle for a proposed modern launch system (Perigee West Company)

The RSX rocket engine was proposed by Rocketdyne as a candidate for powering liquid fueled strapon boosters for the Space Shuttle, replacing the present solid rocket motors. The concept is premised on the ready availability of such an engine, using technologies that are already proven in the highly reliable RS-27 engine used on Delta II, plus existing turbopump packaging arrangements. The only new development for the initial model of the RSX would be a thrust chamber capable of delivering twice the thrust of the RS-27 engine, slightly over 400,000 pounds.

Propellant feed for the new chamber would be delivered by a pair of RS-27 turbopumps, packaged in an arrangement already used for many years on the Atlas launch vehicle. The turbopumps are physically attached to the thrust chamber. Turbine exhausts are equippedwith heat exchangers, allowing production of oxygen gas for pressurizing the oxygen tank. Use of four engines allows reduction of flight acceleration loads by shutting down engines, as appropriate, toward the conclusion of burn time. The same feature provides ability to operate with one engine inoperative over most of the flight. By applying a common engine to the core vehicle and to strapon booster rockets, an opportunity is provided to lower the cost of rocket engines. Quantity production as well as production of engines in a single facility are the two main factors that will help to lower unit cost.

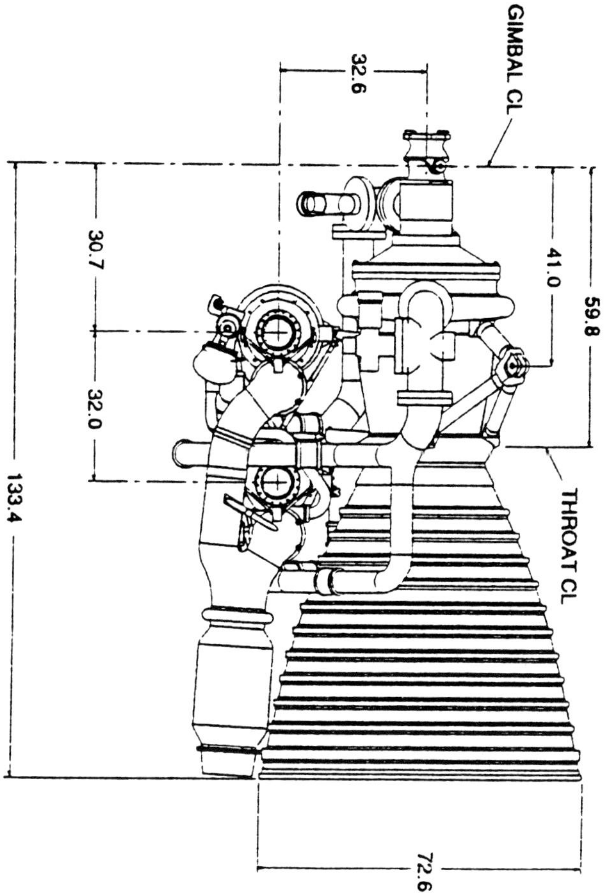

Propellants	LOX/RP-1
Thrust (sea level)	424,000 LB
Thrust (altitude)	480,000 LB
Engine Mixture ratio	2.25
Specific Impulse (sea level	264
Specific Impulse (altitude)	269
Chamber Pressure	775 psia
Weight	4500 LB
Dimensions	134 in. by 72 in. dia.

RSX rocket engine characteristics (Rocketdyne)

The RSX engine concept allows government and industry to proceed with new launch vehicle development with a low investment in propulsion. The engine design is heavily leveraged on existing technology, yet handily does the required job.

An opportunity exists, however, for future improvements that would yield a reduction in engine weight and cost. The first effort would be directed toward development of a new single turbopump to replace the dual system on the original engine. Design of the turbopump would take advantage of technology advancements that were made on the Space Transportation Main Engine (STME), during the NLS program.

Yet another phase of development would aim for a high pressure engine that would yield an additional 30 seconds of specific impulse. This provides for performance growth without changing the physical structure of the launch system. Russian engineers have already achieved this technology in their RD-170 engine.

Adoption of the RSX engine has the immediate benefit of providing a new engine in a short time frame. It also provides a medium for preserving a development engineering operation for some time into the future that might otherwise be lost to the country.

The new launch system works most economically if fitted with a choice of two upper stages, enabling transport of payloads to low earth orbit ranging from 20,000 pounds to 50,000 pounds. One upper stage is a 56,000 pound gross weight high energy vehicle which uses liquid oxygen and liquid hydrogen propellants. This vehicle is a rugged descendant of the General Dynamics developed Centaur upper stage. All technology, including engines, for building it is in hand.

The other stage strives for simplicity to fulfil its role of carrying payloads at the lower end of the weight range. It is a hybrid vehicle weighing 46,000 pounds, and is fueled by liquid oxygen and a solid fuel. This type of stage offers the ultimate in ease of handling and processing, with virtually no safety concerns. Launch processing for a hybrid upper stage is simple, requiring loading of only one propellant.

Development of a hybrid motor will be relatively inexpensive, as demonstrated by American Rocket Company's successful development of several sizes of hybrid motors with private funding.

An important feature of the two upper stages is sharing the same avionics subsystem. This results in important cost savings in hardware development and procurement, and software generation and maintenance.

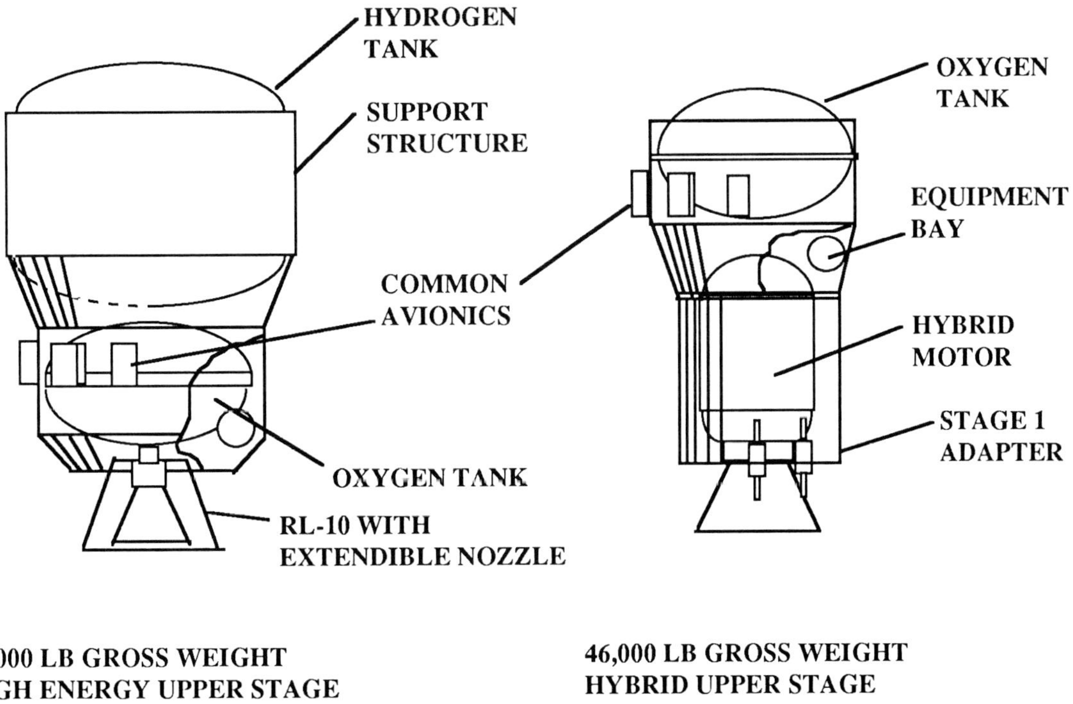

Upper stages designed for the modern launch system (Perigee West Company)

Payloads at the lower end of the weight range can be orbited with the core vehicle, equipped with one or the other of the upper stages. Achievement of the upper weight range requires the addition of strapon boosters.

Strapon boosters can be built at modest cost through an innovative design approach. The entire tank and structure is designed to be fabricated to ASME code. The high factor of safety for code-built hardware eliminates the need for extensive testing before flight. Such structures can be procured from established code weld shops, at a cost of about seven dollars a pound.

Boosters of this type will require no pressurization of the oxygen tank. Gravity head is sufficient for operation of the oxygen turbopump. The thick walls will have sufficient load carrying capability for all launch and flight conditions.

Past launch vehicle studies by TRW have included what have been termed "Big Dumb Boosters." Such boosters were planned to be entirely built to heavy industry standards, including the rocket motor. They were designed to operate at high tank pressures, eliminating the need for turbopumps. The inefficiencies of such boosters are neatly overcome in the present concept, by using the best features of "Big Dumb Booster" and turbopump based vehicles in a single design.

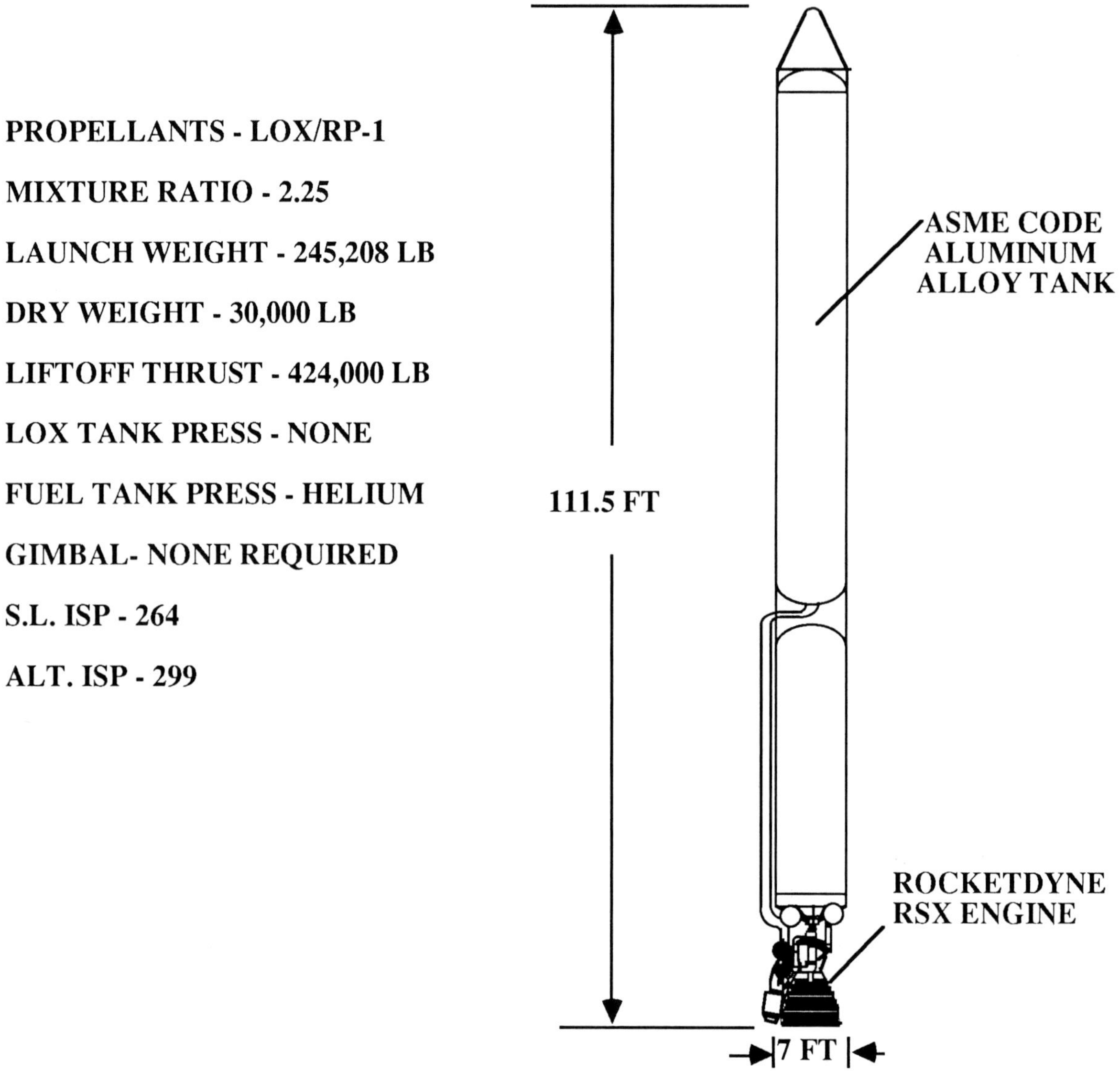

Strapon booster design concept for the modern launch system (Perigee West Company)

Three vehicles comprise the proposed modern launch vehicle family, containing an array of desirable features:

Standard core vehicle
Common engines
Common avionics
Common payload shroud diameter
Shutdown control of all stages
Limited engine-out capability
Operation verified before launch
Low cost upper stage for small payloads
High energy upper stage for large payloads

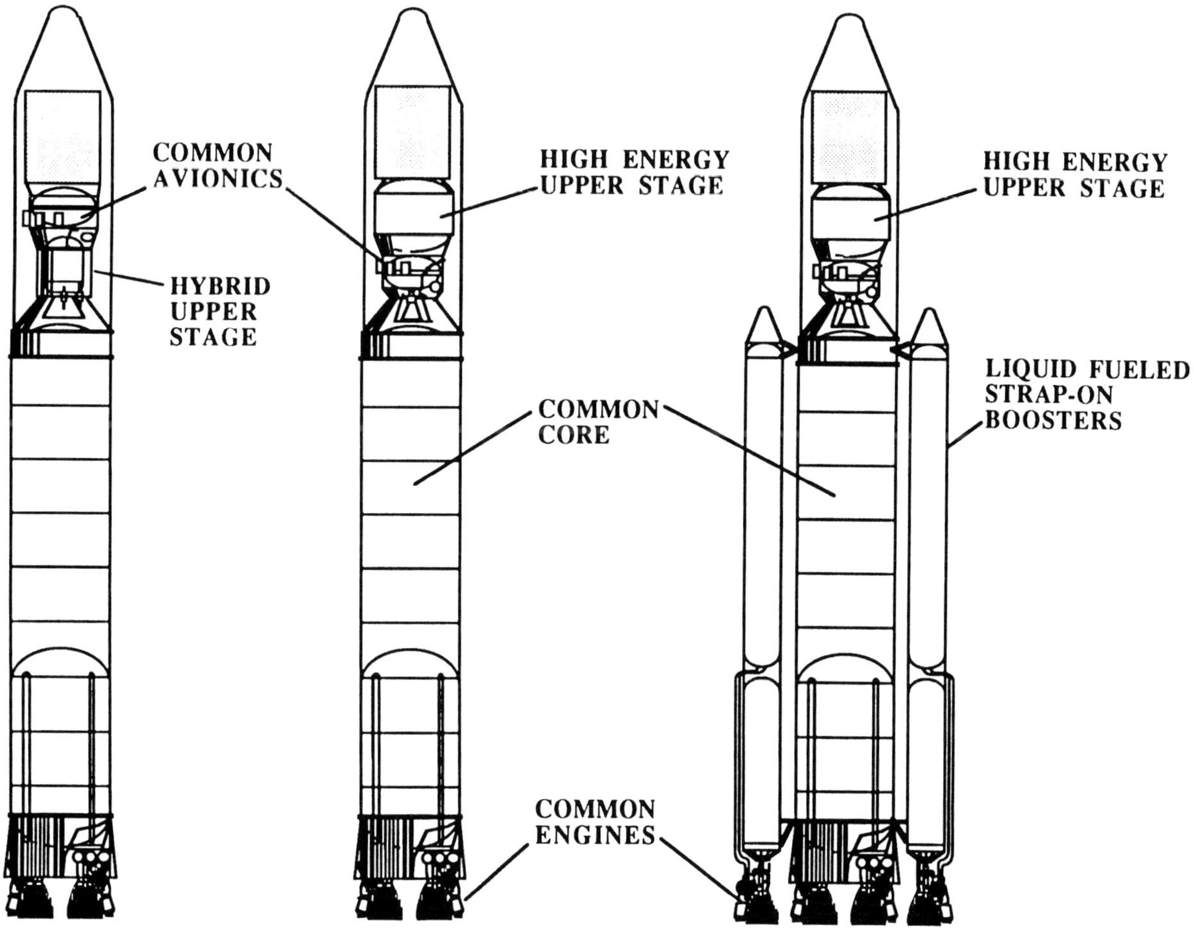

20,000 lb	35,000 lb	52,000 lb	LOW EARTH ORBIT
7500 lb (Kick motor required)	17,000 lb		GEOSYNCH ORBIT TRANSFER
	9200 lb	15,000 lb	GEOSYNCH ORBIT

The three vehicles comprising the modern launch system (Perigee West Company)

10-4 Trouble at the Top

Like the actors in Samuel Beckett's play "Waiting for Godot", the chief executive officers and chairmen of America's leading space launch vehicle corporations pensively await the coming of a government funded program to develop the next generation launch vehicle.

It could be a long, cold wait. Unless a critical national requirement develops, approval of a new launch vehicle program is only a distant possibility. As long as the United States operates with a budget deficit, the absence of a critical requirement will result in a no vote by the Congress that recently decided that the nation cannot really afford the Superconducting Supercollider, the Advanced Solid Rocket Motor, the National Launch System, or the National Aerospace Plane.

No one should believe that the space launch business is a victim of the end of the Cold War. Mission models show growth, particularly in the civilian and commercial sectors. Contraction of the business is largely a consequence of an over dependence on the U.S. government by the leaders of the launch vehicle industry. They know that a firm market exists, serving Defense Department (DoD), civilian and commercial needs for fifty years or more into the future.

That knowledge justifies development of a competitive, all-purpose launch system. But after nearly ten years of study, funded by NASA and DoD, launch vehicle manufacturers do not appear to know what a competitive, all-purpose launch system is. Visions of a reusable SSTO, in a variety of configuration concepts, costing tens of billions of dollars to develop, complicate what is in fact a rather simple picture.

There are three good reasons why a recoverable SSTO does not fit the current launch vehicle requirements scenario. First is the acknowledged high cost of development. The cost of development of the McDonnell Douglas Delta Clipper has been estimated to be as high as $20 billion, when fleet acquisition costs are factored in. The government cannot afford it, and the price is too high for privately funded development.

Next are unknowns regarding whether such a sustem will indeed reduce the cost of launching into space. The old generation of designers failed with the Space Shuttle. The new generation of designers seem to think that it is possible, by taking new design approaches. But the Space Shuttle experience shows that a reusable launch vehicle has a long way to go to even match the economies of an expendable system.

Lastly, it is fundamentally a bad policy to expect a single type of vehicle to handle a broad array of payloads. That is why on Earth we have pickup trucks and eighteen wheelers.

The preceding concept, however, does meet the current requirements scenario, and can be acquired for less money than an auto manufacturer may spend to develop a new model.

The big worry for the U.S. space launch industry is the availability of several foreign launchers to commercial customers. The U.S. government is not likely to underwrite a development effort that enables U.S. industry to compete commercially. The technology is mature. The industry has to do it for itself.

Overall, however, it appears that the managers of the major launch vehicle companies have lost the forward thinking vitality that got them where they are. The prevailing attitude is that if progress is to be made, the government must provide the funding. They could take a lesson from the Boeing company, which recently produced the magnificent 777 aircraft without government funding, employing methods and processes virtually unknown to the rocket industry. But is an airplane as complicated as a rocket? The facts are that it is far more complex.

A full retreat from the space launch system has taken place at General Dynamics, the consequence of a determination by majority stockholders that it would be prudent to get their investments out of the company. The last of the holdings to be sold off was the space systems division. The management of General Dynamics clearly has lost its vision for the future, insofar as space business opportunities are concerned.

McDonnell Douglas management is content with their Delta II launch vehicle, making changes to it in accordance with new government or commercial flight requirements as they develop. The DC-X program has management support if the government is able to fund it. When the 1993 activity on the test article fell short of its goals due to lack of funding, the operation was shut down.

Until its purchase of the General Dynamics space systems division, Martin Marietta's launch operations business has survived almost entirely on government business. An initial unsuccessful venture into commercial launches was quickly aborted. It remains to be seen whether this company can measure up to the demands of the commercial launch world. A second move, far more significant , is the announced merger of Lockheed with Martin Marietta on August 29, 1994. The merger furthers the consolidation of defense business, but holds small promise for commercial enterprise.

Very little in the behavior of the leaders of these companies indicates that they have a vision for the future. If they continue on their present paths, counting on U.S. government funding of a launch system which they should be developing by themselves, they will, ten years from now, still be waiting for Godot.

10-5 Outlook

An atmosphere of uncertainty has descended upon the subject of a modern launch system for the United States. After the latest studies the prevailing mood in Washington is to make do with what we have.. Congressional support for development of a new system is meager. Some support exists for development of a reusable SSTO, but deeper insight will eventually show that the enthusiasm is misplaced.

In August of 1994 President Bill Clinton signed the administration's new space launch policy. It calls for NASA to begin technology development leading to a reusable launch system following the turn of the century. Whether or not emphasis is to be placed on SSTO's is yet to be determined.

Insofar as expendable launch vehicles are concerned, the policy abandons any further consideration of a new system, favoring instead an evolution of existing systems to a greater capability and better economy. To this end, it is expected that the Air Force will compete existing launch systems to arrive determine which of the three, Atlas, Delta or Titan, are best suited to a growth path.

The preceding material describing a modern launch system is well suited to the aims of the new space launch policy, since its design is leveraged on the best performing components of existing systems. It remains to be seen which of the leaders of the two competing companies, McDonnell Douglas or Lockheed-Martin, will demonstrate the courage and insight necessary to carry off a development which will ultimately yield the system that the DoD has sought from the beginning.

11

National Aero-Space Plane

11-1 Wings Into Space

An ongoing dream of rocket engineers since the early 1930's has been to build a combined airplane-like spacecraft that could take off from a runway like an airplane, wing its way into earth orbit and return to earth, to land at an airport like military and commercial aircraft.

The ensuing forty years is strewn with concepts for approaching such a capability, but sensibly, in the face of the formidable technological challenges inherent in such a venture, all projects of this nature received only minimal funding for exploratory studies before being cancelled. In every case the required technologies were far beyond achievability for its particular time period.

Investments were directed instead toward experiments in supersonic and hypersonic flight, and to this purpose a series of rocket aircraft was developed, starting with the Bell X-1 and ending with the highly successful North American X-15. These programs acquired much of the aerodynamic and aerothermal data that was needed for design of the Dyna-Soar, and later the Space Shuttle. The Bell X-1 was the first to break the sound barrier. The Bell X-1 was followed by the Bell X-2 and then by the Douglas Skyrocket. These X planes reached the limit of their usefulness at a peak velocity of about Mach 3. North American was then funded to build the X-15, to explore higher regions of velocity and altitude. By the end of its flight program in October of 1968, the X-15 had reached top speed of Mach 6.7 and a peak altitude of 67 miles.

These rocket powered aircraft were air-launched. They were carried to launch altitude slung under a wing of the B-52 bomber. There was never any thought of launching such a craft from a runway, under its own power.

The problem, of course, is the near physical impossibility of storing enough energy aboard an aircraft-styled rocket to permit it to operate in the manner of conventional aircraft. No combination of known chemical propellants will work. To get around this, designers promoted staged concepts, in which the orbital craft rides piggyback on a larger booster vehicle, both of which operate like an aircraft in a landing mode.

Even if it were possible from an energy standpoint, other factors enter to discourage the picture. The loaded launch vehicles are extremely heavy, requiring extremely massive running gear. Takeoff runways would have to carry much higher loads than currently possible. Emergency landings immediately after takeoff would be impossible. Engineers had to look to alternative schemes for the concept to work.

11-2 The Aerospaceplane

The most ambitious program to develop a reusable single-stage-to-orbit launch vehicle was the *Aerospaceplane,* not to be confused with the more recent *National Aero-Space Plane*, or NASP. The Aerospaceplane project was begun in the late 1950's, under the direction of Weldon Worth, Director of the Aero-Propulsion Laboratory at Wright Field. It was to be a very large winged vehicle capable of taking off and landing on conventional landing fields. The vehicle was to carry a crew of three and had a payload bay measuring 40 feet long by 25 feet wide by ten feet high.

A unique feature of the Aerospaceplane was its propulsion system, which solved the problem of the heavy launch weights that are characteristic of pure rocket vehicles. The propulsion concept was based on recovering most of the oxygen propellant from the atmosphere through in-flight processing of injested air. The rocket fuel was liquid hydrogen, contained in insulated tanks inside the craft.

Designers of the Aerospaceplane devised a liquid air collection system, using the liquid hydrogen carried aboard the vehicle to condense ingested air to a liquid state through an array of heat exchangers. The propellants were then fed into the vehicle's rocket combustors. The initial system was called LACES, for Liquid Air Collection Engine System. The idea was to generate and store sufficient liquid air in the vehicle to enable propulsion operation after LACES ceased ability to power the vehicle into earth orbit.

However, liquid air has a high nitrogen component, which resulted in less than optimum engine performance. Further work by General Dynamics and Marquardt Corporation led to ACES (Air Collection and Enrichment System). This system was to produce oxygen enriched liquid air by separating out significant amounts of nitrogen. A difference in boiling points between nitrogen and oxygen opened this possibility. Marquardt Corporation engineers were able to demonstrate feasibility through experiments performed on a Garrett Corporation heat exchanger.

The program was thrown into confusion with the introduction of the possibility of powering the aerospaceplane with scramjet propulsion. Supersonic combustion ramjets, or scramjets, take over at velocities beyond which ramjets no longer function, about Mach 4.0, and can theoretically operate at Mach numbers up to 18. Scramjet possibilities were introduced in 1958 in a technical paper from NACA Lewis Laboratory. Several laboratories then set about trying to solve the problems of supersonic combustion. Aerospaceplane and scramjet propulsion seemed to be ideally suited for each other, giving momentum to the idea that scramjet based propulsion was a more attractive way of creating an Aerospaceplane.

Lead by chief engineer Alexander Kartveli, Republic Aviation Corporation jumped into the arena with proposed scramjet powered vehicles. Their final configuration was a swept wing arrangement with an underbelly scramjet installation, a harbinger of designs that would be featured in NASP almost 30 years later. A single aft-mounted rocket engine provided power when the scramjet was no longer operative, and jet engines were used for takeoff, cross-range flight and landing.

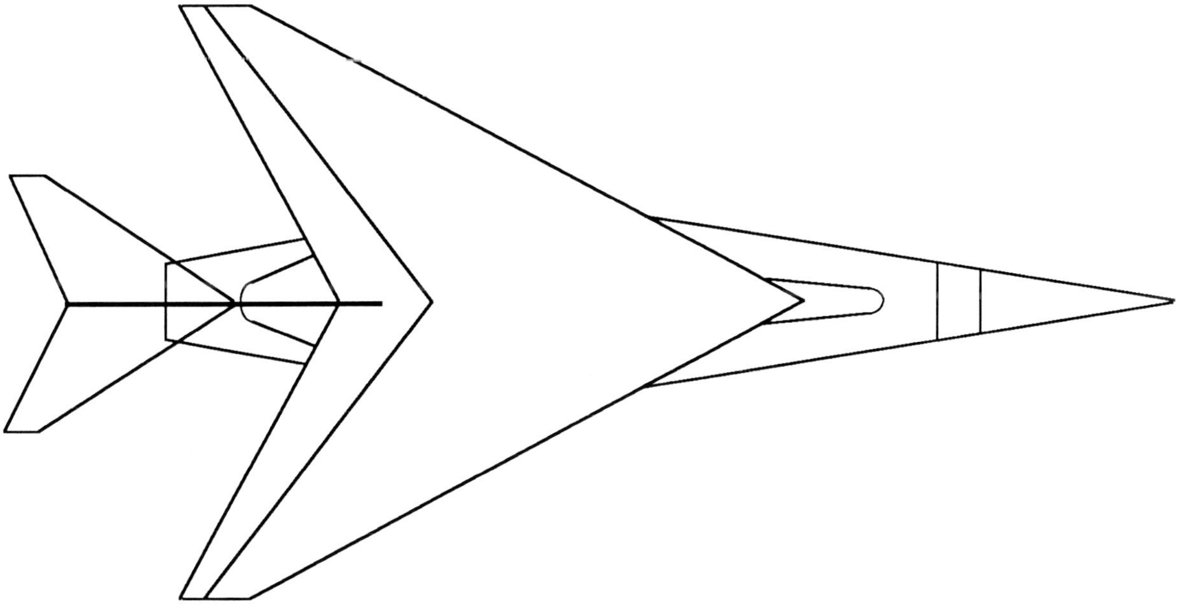

Top view

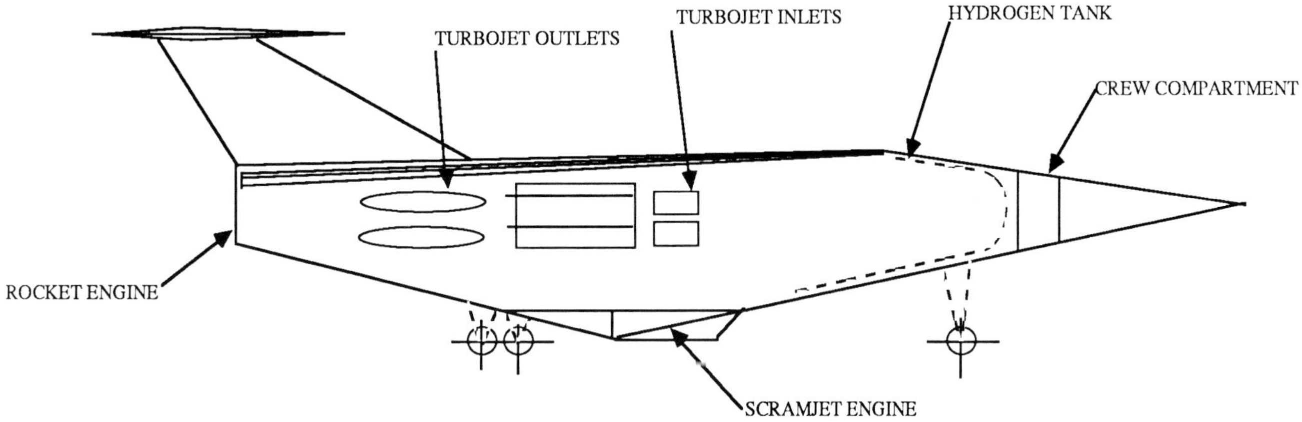

Republic Aviation's concept for a scramjet powered Aerospaceplane, circa 1962

Although technical problems with the Aerospaceplane were acknowledged to be formidable, the Air Force remained optimistic about the prospects of acquiring a reusable launch system for a variety of military missions. The Aerospaceplane appeared to be a good way to transport material from point-to-point on Earth very rapidly, should requirements develop in that direction. By 1962, seven contractors, Boeing, Douglas Aircraft, General Dynamics, Goodyear, Lockheed, North American and Republic were deeply involved in Aerospaceplane studies. In June of 1963 three contractors, General Dynamics, Douglas and North American were awarded $500,000 studies to undertake detailed development planning. Other contracts were underway to continue technical research on materials, oxygen-enrichment equipment, scramjets and other subsystems.

The Space Advisory Board (SAB) was concerned, however, that too much effort was being spent on operational aspects of the Aerospaceplane and too little on research and development. This lead to a conclusion by the SAB in October of 1963 that the nation's technology base was not up to the task of developing a single-stage-to-orbit reusable space transporter. The Aerospaceplane project was cancelled when congress deleted all fiscal year 1964 funding for the project.

Work continued toward developing the various technologies required for a single-stage-to-orbit, but since it still remains classified, little is publicly known of it. Much of it would presumably have found application when the government elected again to undertake development of the NASP during the late 1980's.

The late 1950's and early 1960's was a period during which many new ideas for space systems surfaced, energized by the adversary situation with Soviet Russia brought about by the cold war, and the race for superiority in space that began with Sputnik. Dyna-Soar was actively being developed and considerable activity was underway on small lifting body designs, that would permit one or two crewmen to orbit and return to an unpowered landing at an airport. Except for Dyna-Soar, none of the programs was heavily funded. There was a willingness to fund studies of new and innovative ideas, if only to see where the work would lead. When appropriate, it was a simple matter to terminate funding if it was judged that no further work was desirable.

Total funding for Aerospaceplane was probably under $5 million. For that expenditure, the doors were opened to a wide array of materials and subsystem advancements, many of which in one form or another have found their way into existing flight systems.

11-3 NASP - The National Aero-Space Plane

Interesting overseas developments during the 1980's lead to a resurgence of Aerospaceplane activity in the United States, only this time the initiative was named National Aero-Space Plane, or NASP, as it is commonly termed. NASP had the same initial objectives, namely to be capable of taking off from an aircraft runway with a payload, ascend to earth-orbit and return to an airfield, using operations practices that are common to commercial aircraft.

The Europeans were working toward development of their own manned vehicle, Hermes, which was to be launched atop Ariane. The French space agency CNES outlined the program during a 1982 symposium. Hermes seemed to many engineers in the United Kingdom to offer little in terms of real progress. It appeared to be something like a repeat of the ill-starred U.S. Dyna-Soar effort twenty years previous. The CNES/European Space Agency plan spurred British inventor Alan Bond, currently the managing director of Reaction Engines Ltd, into a reexamination of advanced launchers. That effort grew to a program conducted jointly between British Aerospace and Rolls Royce to develop a launch vehicle called HOTOL. This program, and in particular the innovations in the HOTOL engine, was an important factor in the start-up of the U.S. activities.

Candidate propulsion systems at that time were turboramjets, liquid air cycle, or LACE powered vehicles as had previously been studied under the Aerospaceplane program, and supersonic combustion ramjets, or scramjets.

Turboramjets can operate to about Mach 5, after which the craft would have to resort to other propulsion to reach the desired Mach 25 needed for orbiting the Earth. The devices are heavy, and would impose an unwanted weight burden, since they would be transported into orbit with the vehicle.

By outward appearance, scramjets seem to be very simple devices. Were they designed for one speed, they would indeed be simple. But in a single-stage-to-orbit application the scramjet is faced with a constantly changing inlet air flow condition. The engine design therefore has to respond to the changing conditions through adjustment of the inlet aperture, as well as adjustments within the mixing and combustion sections of the engine. Scramjets therefore become variable geometry devices.

The upper limit of scramjet performance was at first believed to match the velocity necessary to orbit a spacecraft. Initial enthusiasm for the concept of a scramjet driven single-stage-to-orbit was based in part on this premise. Now engineers admit that the upper limit is in the region of Mach

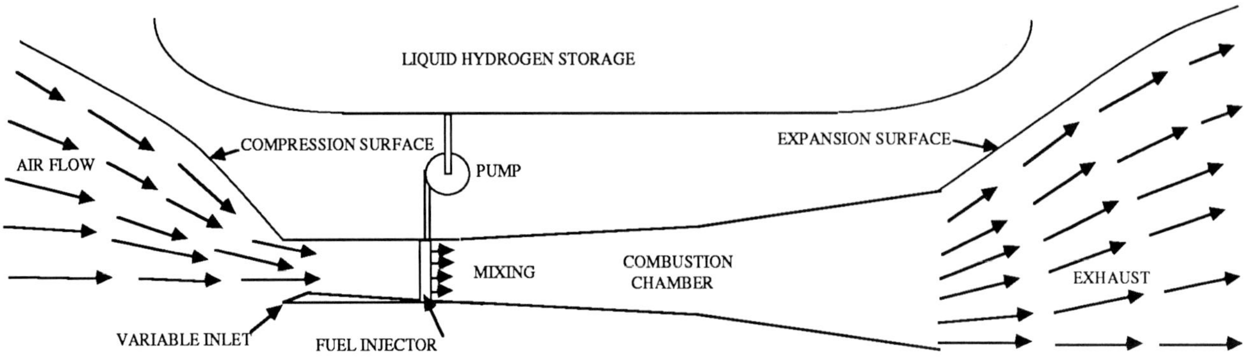

Diagram of a scramjet engine

16, possibly as high as Mach 18. The remainder of the velocity needed to attain orbit is conceded to be obtainable only by rocket thrust.

A key problem confronting designers of an air breathing SSTO is how to provide thrust to the craft over ranges where a scramjet is inoperable, from zero velocity to about Mach 5, and from Mach 16 to Mach 25.

During Alan Bond's work on the engine system for HOTOL it was concluded at an early stage that the cost, technical risk and extreme complexity of a scramjet device showed low promise for low cost transport to low-earth orbit. It was difficult to show that scramjets possessed advantage over other types of propulsion systems and the cost of engineering was believed from the beginning to be an overwhelming factor in the selection of alternative systems. While the scramjet might prove to be a good device for high mach number cruise operation, it was believed to be a marginal choice for an accelerator, as is required for a space booster.

Bond and his crew turned their attention to the liquid air cycle engines that had been investigated more than twenty years previous. But the new wrinkle was to create an engine system that operated with air at near liquid air temperatures instead of liquid. This enabled them to devise an engine that would operate as a rocket, rather than as a scramjet. A turbine driven air compressor was used to compress the incoming air to pressures sufficiently high for operating a rocket engine. Without going into the complexities of the engine, its capabilities were such that the HOTOL vehicle could take off starting at zero velocity from an airfield and accelerate under rocket power, using cooled compressed air for oxidizer, to about Mach 5. At this point the air intake is shut off and the vehicle accelerates to orbital velocity with liquid oxygen being fed to the rocket engine from tanks aboard the vehicle.

HOTOL received much publicity, but eventually the British govern-

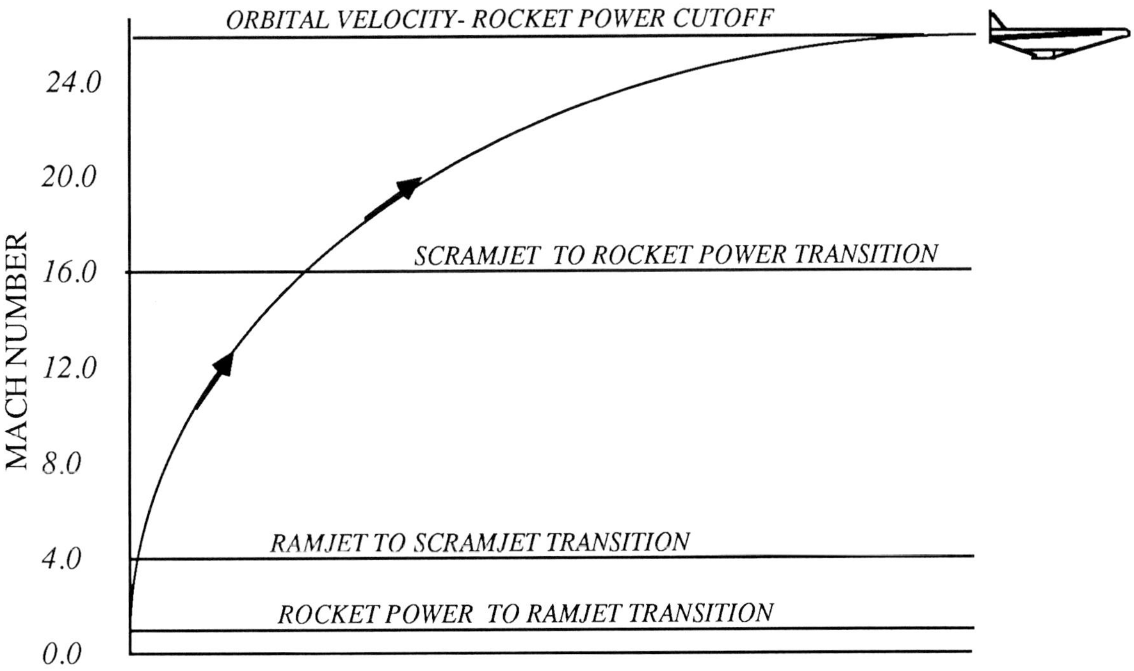

Flight ascent path of an air-breathing SSTO showing types of power systems required over the range of velocity gain.

ment decided to cease funding, upon assessing the bleak projections for payback by the HOTOL venture. In 1987 the program was placed on hold at British Aerospace and at Rolls Royce, where Alan Bond was heading the engine work. By this time an ambitious effort on NASP was under way in the United States, headed by Wright Aeronautical Laboratories, with participation by several leading U.S. space manufacturers.

In 1990 additional HOTOL activities by British Aerospace became public knowledge when the Russians revealed that wind tunnel tests had confirmed that it would be feasible to launch an interim version, using the Antonov AN-225 as a carrier vehicle. The new HOTOL would use a Russian rocket engine in place of an air breathing engine and could place a 16,000 pound payload into a 180 mile altitude equatorial orbit.

Rocket powered version of HOTOL, launched from an AN-225

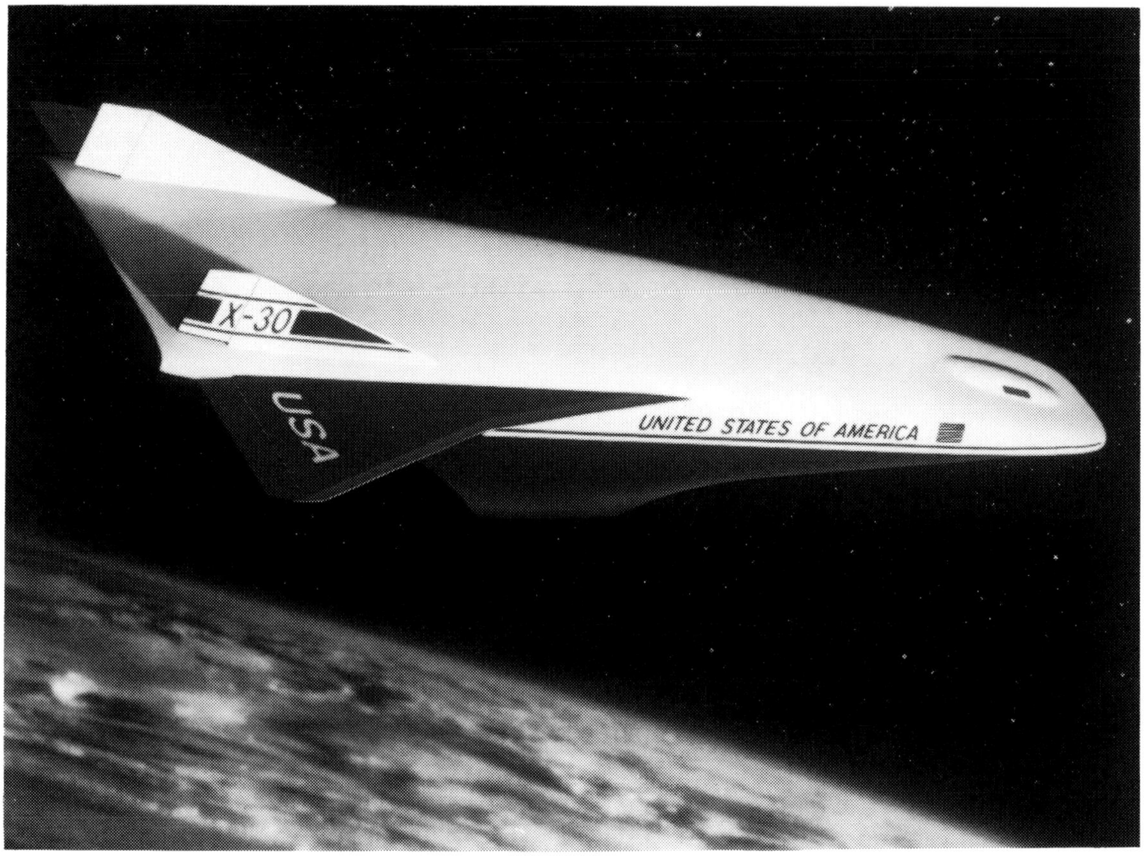

Artist's concept of the U.S. National Aero-Space Plane

The National Aero-Space Plane is a classified project, so details of the program remain obscured. Some conclusions can be made from historical developments that indicate changes in objectives for the program.

Initially the objective for NASP was to accomplish what the Space Shuttle failed to do, i.e. provide an economical way to transport payloads to low-earth orbit, following an operations concept closely resembling that being used in the commercial aircraft industry. While Alan Bond's engine concept appeared to be of interest to the project's early participants, the idea of a combined ramjet/scramjet engine appears to have gained favor. Somewhat later, it appears that NASP officials and engineers arrived at the same conclusion as Bond had reached regarding the impracticability of such devices in an acceleration application, which is all that launch vehicles do. When this happened, enthusiasm for the orbital craft waned, and the idea of a very fast intercontinental transport took hold.

By this time, the program had received President Reagan's attention and approval. The President announced that the nation was going ahead with development of the "Orient Express," a hypersonic transporter that would permit flights to the far East in a matter of an hour and a half or so. NASP now became a Space Plane only in the sense that it neared space as it followed its trajectory toward the landing point.

The "Orient Express" idea eventually gave way to a development program for an experimental hypersonic aircraft dubbed the X-30. Several years of work at respectable funding levels were performed by industry teams to try to solve some of the many difficult problems inherent in the hypersonic vehicle.

The X-30 faces an array of technological problems which are difficult to solve using existing tools. Wind tunnels do not exist that can produce the high velocities needed to prove operability of scramjets. At best, engineers must resort to shock tubes which yield only milliseconds worth of test data. The dilemma has spawned large advances in analytical capability in computational fluid dynamics, through which attempts are made to predict behavior of airflow and combustion flow covering various engine geometries. Unlike jet and rocket engines, which can be tested and qualified in ground testing facilities, actual testing of a scramjet engine can only be accomplished by flight testing, which is very expensive. A development model might be tested by this method, but flight qualification, which involves extended testing, is an entirely different story.

Participants in the ramjet/scramjet engine research and development studies include the principal jet engine manufacturers in the United States, General Electric and Pratt & Whitney, as well as at least one rocket engine company, Rocketdyne Division of Rockwell International.

The aircraft itself presents monumental thermostructural and aerodynamic challenges arising in part from prolonged atmospheric heating during its acceleration phase, and unknowns with respect to boundary layer conditions at hypersonic speeds. Engineers and scientists must learn what parts of the craft are subject to turbulent flow at the boundary layer and take appropriate measures to improve the structure at those points so they can withstand the heat.

If powered by a ramjet/scramjet engine combination, the aircraft is literally a flying tank of liquid hydrogen, at a temperature of minus 460 degrees F, closely coupled to an external shell that varies in temperature over thousands of degrees F. Materials studies performed in the course of the NASP program have advanced the idea of using hydrogen tanks made of graphite composite instead of metal, for added strength and weight savings. If rocket power is included in the propulsion system, as would be the case for an orbital craft, the plane's fuselage would also contain a liquid oxygen tank.

Whether or not this is a comfortable situation for the crew members and passengers, who would be seated forward of the tanks, is a matter for conjecture. Only a few years have passed since NASA rejected flying the hydrogen fueled Centaur in the cargo bay of the Space Shuttle, primarily because it was deemed too hazardous for the crew members.

11-4 Outlook

Like a recurrent dream, the idea of flying into space in an aircraft type of vehicle - in a single stage - keeps popping up. Each time significant progress is made toward establishing the technological base that can make this dream possible. Perhaps on the next cycle it will happen. At least it is now recognized as a formidable technological task. Whether there is a payoff is at present a complete unknown. The single data point presently available, the Space Shuttle, has demonstrated that reusable space launch vehicles are far more expensive to operate than expendable rockets. A present-day challenge, before large investments are made again in this direction, is to prove that reusable launch vehicles are a panacea for low cost earth-to-orbit transport.

The X-30 program has hit on lean times due to the U.S. budget problem, but remains on the books for technology funding at a low level. If the far term aim of the X-30 program is indeed to make possible intercontinental trips at hypersonic speed, then saner minds have probably questioned the requirement, at least for our time period. The Concord experience has failed to spur commercial aircraft development even at supersonic speeds, instead revealing that such ventures are an economically losing proposition. A hypersonic aircraft, as presently conceived, would be an uncomfortable flying situation for most passengers.

Curiously, developments in other quarters may have already bypassed the X-30 program. This can only be surmised, but it is difficult to accept the Air Force's retirement of the SR-71 fleet unless it had already been replaced with something of far greater performance. One can envision that turboramjet development has progressed far enough to make possible an aircraft that can take off under its own power and reach speeds up to Mach 5 or Mach 6. If this is so, then the value of the X-30 program would rest primarily on benefits to be gained from pursuit of scramjet powered flight. As for the future of the "Orient Express," the power plants that can make possible flights to Japan in an hour or so may already be in hand.

Meanwhile, back in England, Alan Bond is hard at work at his new company, Reaction Engines Ltd., defining a new single-stage-to -orbit launch vehicle called Skylon, built on lessons learned from HOTOL and incorporating new ideas on how to build an air breathing rocket engine. The new "Sabre " engine is purported to have better performance than the engine originally devised for HOTOL. It operates as an air-breather up to Mach 5 and then converts to pure rocket operation for the balance of the flight. Where all this will lead in the absence of British government support is questionable. Perhaps in the long run an undertaking of this nature should be international in scope, since in the broad view it should serve world needs, as opposed to the needs of a single country.

12

Ultra Large Launch Vehicles - ULLV's

12-1 Thinking Big

Repeatedly, since the advent of space rocketry in the United States, engineers have turned to space projects that require rocket lift capability up to three orders of magnitude greater than derivatives of the ICBM program were able to launch. There is hardly a limit to the weight that can be orbited. The Empire State Building could be launched into orbit provided a large enough rocket were placed under it.

At this time, Saturn V, built during the 1960's for the Apollo program, remains the largest launch vehicle ever built in the U.S. Saturn V was capable of launching 250,000 pounds of payload into low-earth orbit, but was phased out when the Apollo program ended. Its five F-1 booster engines generated a total thrust of 7.5 million pounds. Saturn V was used for a total of fourteen flights.

The next highest lift capability currently resides in the Space Shuttle, which was designed to lift 65,000 pounds, but in fact performs at approximately 10,000 pounds short of that weight. The Shuttle program has accomplished over fifty flights.

The highest lift capability in expendable launch vehicles is claimed by Titan IV, at 39,000 pounds. This vehicle's lift capability will grow to 50,000 pounds when the up-rated solid rocket motors are installed. Titan IV has flown a half dozen times but has a long history in different versions of Titan III, which could orbit about 30,000 pounds.

The real Titans of the launch vehicle world aim for much heavier capability. Payload weight targets ranging between 400,000 pounds and a million pounds are common among the numerous launch vehicle studies that have examined exploration of space on a grand scale.

The Orion scheme for launching multi-kiloton payloads, promoted during the early 1960's by a group of engineers at General Atomics, employed atomic bombs for propulsion. The bombs were programmed to explode serially against a large plate that was attached by springs to the payload above it. Such schemes are of course preposterous in current thinking, as atmospheric nuclear detonations are prohibited.

ULLV's are simply bigger versions of their small payload cousins. More propellants, more thrust, bigger structures. They are controlled in the same manner by identical avionics systems. Although most designs show them to be equally complex, there are some sizing aspects of heavy lifters that make it possible for them to be the simplest rockets of all.

12-2 Who Needs ULLV's?

No one needs ultralarge launch vehicles, at least not at present. But in the course of blue-sky studies that are constantly underway regarding new ventures in space, ideas occasionally surface that identify requirements for lift capabilities far beyond those presently in use. More often than not, these studies are under private authorship and address widely divergent ideas regarding mankind's future presence in space.

NOVA and Post-NOVA studies sponsored by NASA during the early 1960's addressed launch requirements for an ambitious space exploration scenario that was to follow the Apollo program. Heavy-lift launch vehicles were needed for lifting large quantities of propellant into orbit for fueling reusable orbit-transfer vehicles. They were needed to launch orbit-transfer vehicles with their initial loads of propellants. Large shipments of materials for construction of spacecraft, Lunar and Mars basing equipment, and provisioning also entered into the picture. None of this came to pass, however, as manned space initiatives came to a halt with the conclusion of the Apollo program.

A new requirement for heavy lift launch vehicles arose during the 1970's when the U.S. Department of Energy and NASA jointly undertook studies of space power systems (SPS). Rationale for this exploratory effort was that solar energy exists as a constant source of power to replace rapidly depleting earth-based fossil fuel sources.

The plan, based on original work by Peter Glaser of Arthur D. Little, Inc., involved construction of very large orbiting spacecraft consisting of arrays of solar cells for direct conversion of solar energy to electrical power. The plan envisioned beaming the power to Earth via microwave or laser medium. If microwave, the energy was to be collected by ground antennas and reconverted to electrical power, which was then to be fed into distributing power grids.

This project required launch vehicles with lift capability approximating a million pounds, and an ability to operate at high launch rates. The Boeing company, which was a principal player in the SPS studies, prepared preliminary designs for a reusable launch vehicle that was capable of lifting a million pounds to orbit and returning to a landing pond in the vicinity of the launch site.

The SPS never went beyond paper studies, but the ideas linger as a future energy generating possibility. More recent proposals by independent investigators have suggested that the moon is a better location for basing a solar power system. Proponents argue that a large part of the materials needed for an SPS can be produced from raw materials on the moon, thus reducing greatly the amount of material that would have to be launched from the Earth. At present, however, there appears to be no interest whatever within either NASA or the Department of Energy to fund further investigations of Solar Power Systems.

The most recent interest in ULLV's derives from former President Bush's announcement of the Space Exploration Initiative (SEI). Were such a program to be pursued on an ambitious scale, ultralarge launch vehicles would prevail as the most economical way to carry hardware and propellants into Earth orbit.

The SEI, of course, was a short-lived program. It lacked a strength of conviction by its originators, was half -heartedly supported by NASA, and tripped over itself by the publicizing of ill-conceived ideas for specific missions in advance of the appearance of a reasonable long range plan. The program received no support in Congress and expired early as a result of lack of interest.

In September of 1991, while the SEI was still active, NASA Marshall Space Flight Center again surfaced an interest in design studies relating to ULLV's. Of four studies funded under a National Research Announcement (NRA 91-MSFC), one was directed toward investigations of ULLV's with payload capabilities ranging from 400,000 pounds to 600,000 pounds to low-earth orbit. The study contract was awarded to Lockheed Company. But as interest in SEI subsided, the contract was redirected to address launch vehicles of much smaller payload capacity.

Until now, three types of physical arrangements have characterized designs for ULLV's. One type is in-line construction, in which the vehicle's stages are stacked one upon another. Saturn V is an example of this type of construction. The second type of construction permits shorter vehicles than in-line, by arranging boost and second stages in parallel. Titan IV is an example of parallel construction. In the third type, the entire rocket consists of a single stage, which by itself can lift very heavy payloads. More about this in Section 12-4.

A difficulty with ULLV's, if constructed in-line, is the extreme height of the assembled vehicle with its payload. One million pound payload launch vehicles that have been studied by NASA in the past have been as tall as 500 feet in a two-stage configuration.

The tallest, perhaps, was Sea Dragon, a vehicle so huge that designers had to adopt shipyard processes for construction. When finally completed, the launch vehicle was to be towed out to sea, turned to a vertical position, loaded with propellants while floating, and launched from the water. No launch pad was required. Sea Dragon was investigated among a number of other concepts by NASA and industry in 1963, in preparation for what was known as the Post-NOVA program Two vehicles were to be launched per month, with payloads of one million pounds.

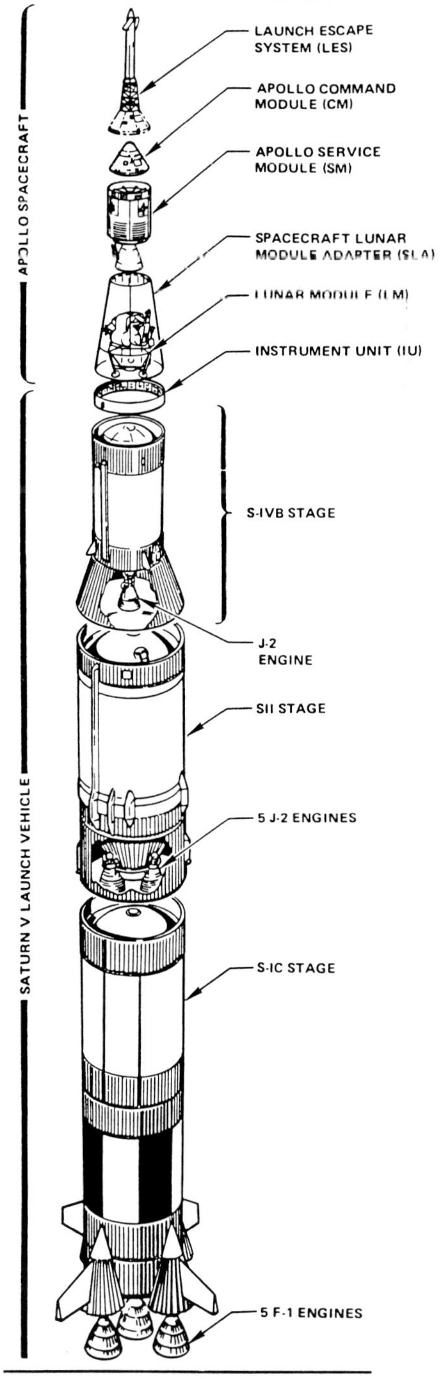

Saturn V is an example of an in-line launch vehicle. Saturn V had a LOX/RP-1 boost stage and two LOX /hydrogen upper stages.

12-3 Cluster ULLV's

Most recent thinking regarding ULLV's has favored parallel configurations, or a clustering of rockets to achieve desired lift capability, following the practices already in place on smaller launch vehicles such as Titan IV and Space Shuttle. In these arrangements, a core rocket comprises the second stage of the vehicle and two or more booster rockets are "strapped" to the core vehicle. When the booster rockets burn out, small, lateral-firing solid rockets are used to separate them from the core vehicle.

The booster rockets can be either solid propellant or liquid propellant devices, or they can even be hybrid rockets. Liquid propellant boosters can be either pressure-fed, by pressurizing the main tanks, or they may use turbine powered pumps to feed propellants to the main engine. Pump-fed boosters yield better performance than solid boosters.

A pressure-fed booster is likely to be powered by a single large thrust chamber, while a pump-fed may have several thrusters. Pump-fed boosters also perform better than pressure-fed due to lighter overall weight and more efficient engines. A hybrid rocket, which uses liquid oxidizer and solid fuel, can yield performance capability comparable to liquid fueled rockets if the oxidizer is pump-fed.

Besides being better performers, liquid propellant boosters are preferred over solid propellant for several reasons. They are easier to assemble and handle than solid rockets. Being empty until loaded on the launch pad, they are relatively light in weight and more easily transported. Solid propellant boosters are extremely heavy and require special safety precautions during launch vehicle assembly and launch processing. Liquid propellant boosters are also preferred because they can be shut down on command. This allows launch personnel to confirm satisfactory engine operation before launch release. Flight programmers also have the flexibility of scheduling engine shutdown at a prescribed time, as opposed to solid rockets being required to operate until burnout. Hybrid rockets possess many of the attributes of liquid-fueled rockets, besides being inherently safe. But they are still in an early phase of development.

Two rocket engine candidates already exist for future ULLV boosters based on cluster arrangements; a modernized version of the Rocketdyne F-1 engine that was used on Saturn V, and the Russian RD-170 Energia boost engine. The latter is presently licensed to the Pratt & Whitney Company for distribution in the United States. One candidate engine, the Space Shuttle main engine (SSME), is already available for second stage use, and the Space Transportation Main Engine (STME), originally intended for the NLS, has at least begun development by NASA. The SSME however, at more than $38 million a copy, is far too costly for use on an expendable launch vehicle.

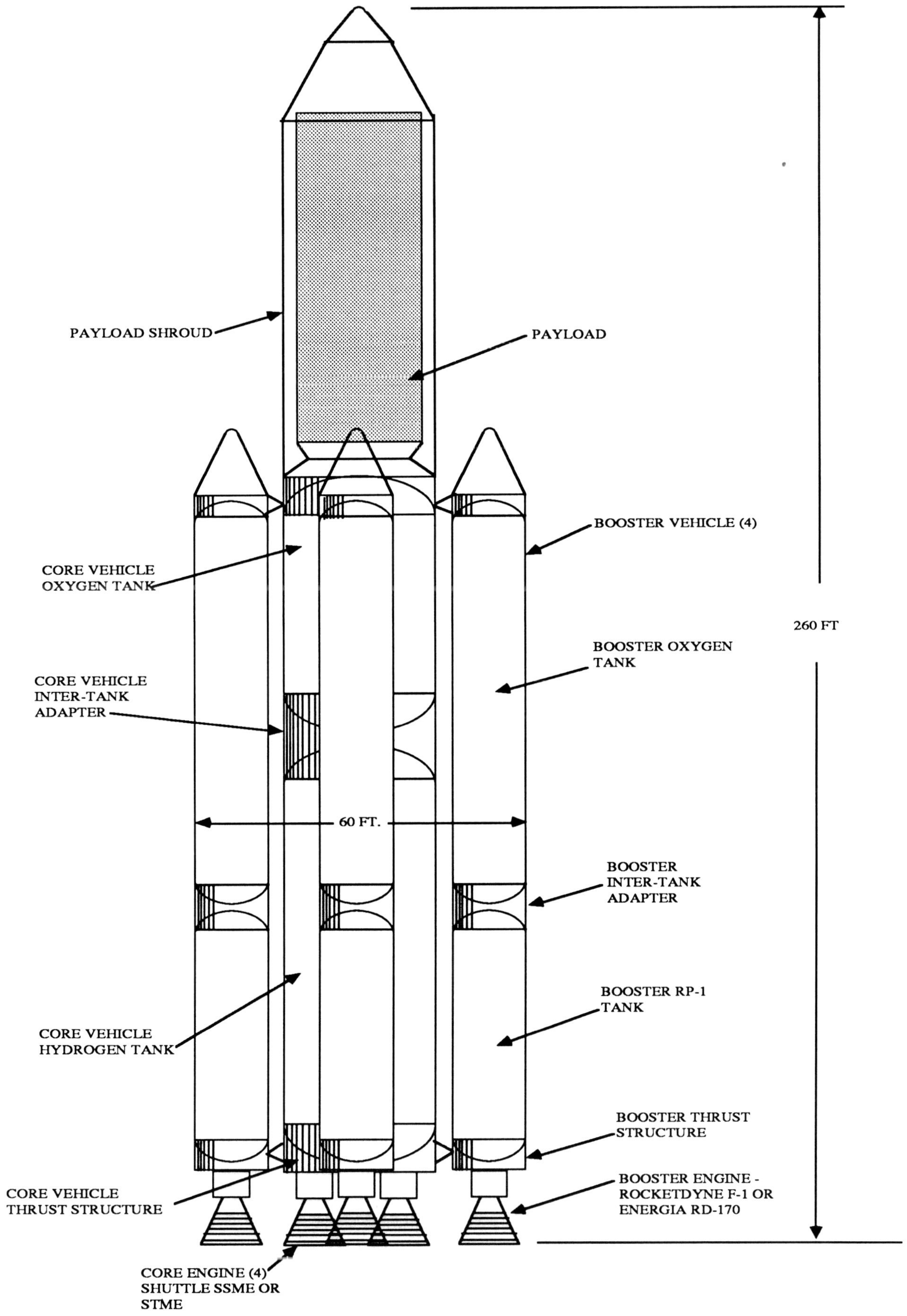

A ULLV concept based on multiple boosters in a parallel arangement.

Rocketdyne's F-1 engine has received considerable interest with respect to possible application in future launch vehicles. Until now there has been a reluctance to consider it, since the engines are more than twenty years out of production. Recent work by Rocketdyne, however, has pointed the way toward reacquiring the engine, designated F-1A, while at the same time incorporating design and processing improvements to produce an improved product.

F-1 A Engine Features

Sea Level Thrust	1,800 K lb
Vacuum Thrust	2,050 K lb
Sea Level Isp	271.0
Vacuum Isp	304.2
Chamber Pressure (psi)	1,161
Mixture Ratio	2.27
Expansion Ratio	16.1
Length x Diameter (in)	220.4 x 143.5
Weight (lb)	19,000

F-1 A Engine Improvements

Oxidizer inducer diameter increased
Decreased oxidizer shaft bore
Increased head 6+6 LOX impeller
Increased Head 6+6 fuel impeller
30 inch turbine
Improved heat exchanger
Strengthened turbineexhaust manifold
Strengthened gimbal seat & cross block
Strengthened high pressure ducts
Strengthened thrust chambers
Strengthened nozzle extension

The F-1 A, modernized version of the F-1 engine has been proposed for future heavy lift launch vehicles (Rocketdyne)

12-4 The Euclid System - a Single-Stage-to-Orbit ULLV

It is clear that cluster ULLV's will be very expensive and complex devices, since several complete rockets are required for a single launch vehicle. As compared with the previous cluster concept, a SSTO will require 60 % fewer subsystems. This reduction in complexity inevitably begs the question: Why can't the job be done with a single stage? The answer is that it can.

In Chapter 8 the McDonnell Douglas reusable Delta Clipper was described. This vehicle is designed to operate as a SSTO. Doubts were expressed, however, regarding capability of the design to operate successfully as a payload carrier. This stems from many prior attempts to achieve such a capability, only to find that the vehicles as finally designed were very marginal in lift capability. The McDonnell Douglas concept attempts to correct this through the use of lightweight materials, but it has taken on weight burdens that are a consequence of the desire to make the vehicle recoverable and reusable. The trend toward operation as advertised is, however, established by the vehicle's large size.

Liquid oxygen (LOX) and liquid hydrogen are the only commonly available propellants that offer prospect of achieving single-stage-to-orbit capability. This combination yields the highest performance in terms of specific impulse, i.e. the thrust obtainable per pound of propellant burned per second. The specific impulse of LOX/hydrogen engines is about 450 whereas the value for LOX/kerosene engines is closer to 300. But, because liquid hydrogen is a very low-density fluid, containment vessels are very large. This has a negative affect on the weight of the whole vehicle due to greater tank weight as well as added structural weight required to support the tank.

If one were to take the physical size of a LOX/hydrogen rocket to extreme, surprising things happen. As the launch vehicle is made larger, it becomes more efficient. In very large sizes, single-stage-to-orbit launch vehicle are efficient payload carriers. This is because the vehicle empty weight increases roughly as a square function, whereas its capacity for propellant increases as a cube function. The net result is a vehicle with a very high propellant mass fraction, which means it can perform well as a single-stage-to-orbit device.

Manufacturing the major components of a SSTO, which comprise the tanks and structure, involves fabrication of fairly large objects. Technology for operations of this nature has already been proven in at least one instance. During the 1970's, General Dynamics operated a production line for spherical aluminum tanks, 120 feet in diameter. The tanks were produced at a rate of one every eighteen days and shipped via barge to the shipyard where they were assembled into liquid natural gas ships.

The accompanying line illustration of the " Euclid" ULLV is capable of launching 600,000 pounds into low-earth orbit. It can be used for smaller payloads by reducing the number of engines and loading the tanks to partial capacity.

The structure and tanks are made of the new "Weldalite" aluminum-lithium alloys that have recently been proven to be highly suitable for producing lightweight rocket components. These alloys, developed by Martin Marietta corporation, are presently planned for production of light weight Space Shuttle external tanks.

The propellant tanks are spherical and are supported by a strut system that attaches to the outer structure. The large size of the hydrogen tank gives visual impact to the affect of this low density propellant on the physical dimensions of the vehicle. At that, the total weight of oxygen in the oxygen tank is five times that of the tanked hydrogen. The propellant feed line from the oxygen tank runs through the hydrogen tank to a distribution header, from which it is fed to the engines.

The propellant tanks are pressurized by warm oxygen and hydrogen gas, respectively, drawn from heat exchangers on the rocket engines.

Euclid's rocket power is provided by 20 Space Transportation Main Engines. The STME has been under development as part of the now dormant National Launch System that was recently under joint study by the DoD and NASA. The engines swivel radially on their mounts for steering. The launch vehicle lifts off at about 1.5 g and can be programmed for moderate acceleration levels by the simple process of shutting off engines at appropriate times in the launch trajectory.

The payload fairing can protect an interior payload, or it may be part of a payload, for example a shielded spacecraft or a propellant carrier.

Assembly of the Euclid ULLV is planned to take place on launch barges, which would be dry-docked inside large buildings during the assembly process. The major vehicle elements would be hoisted into place and fastened together. The ULLV is secured to launch pedestals and stands over an opening in the barge through which engine exhaust is ported at liftoff.

A single-stage-to-orbit ULLV requires no new technology and would be far more economical to build than its cluster counterpart. Future space initiatives that call for very heavy lift capability will in all probability enter concepts like the Euclid ULLV into design trade studies. If an engine is already available as a result of development efforts related to the Air Force's Spacelifter program, a system like this could be produced with far less effort than was required to design and build Saturn V.

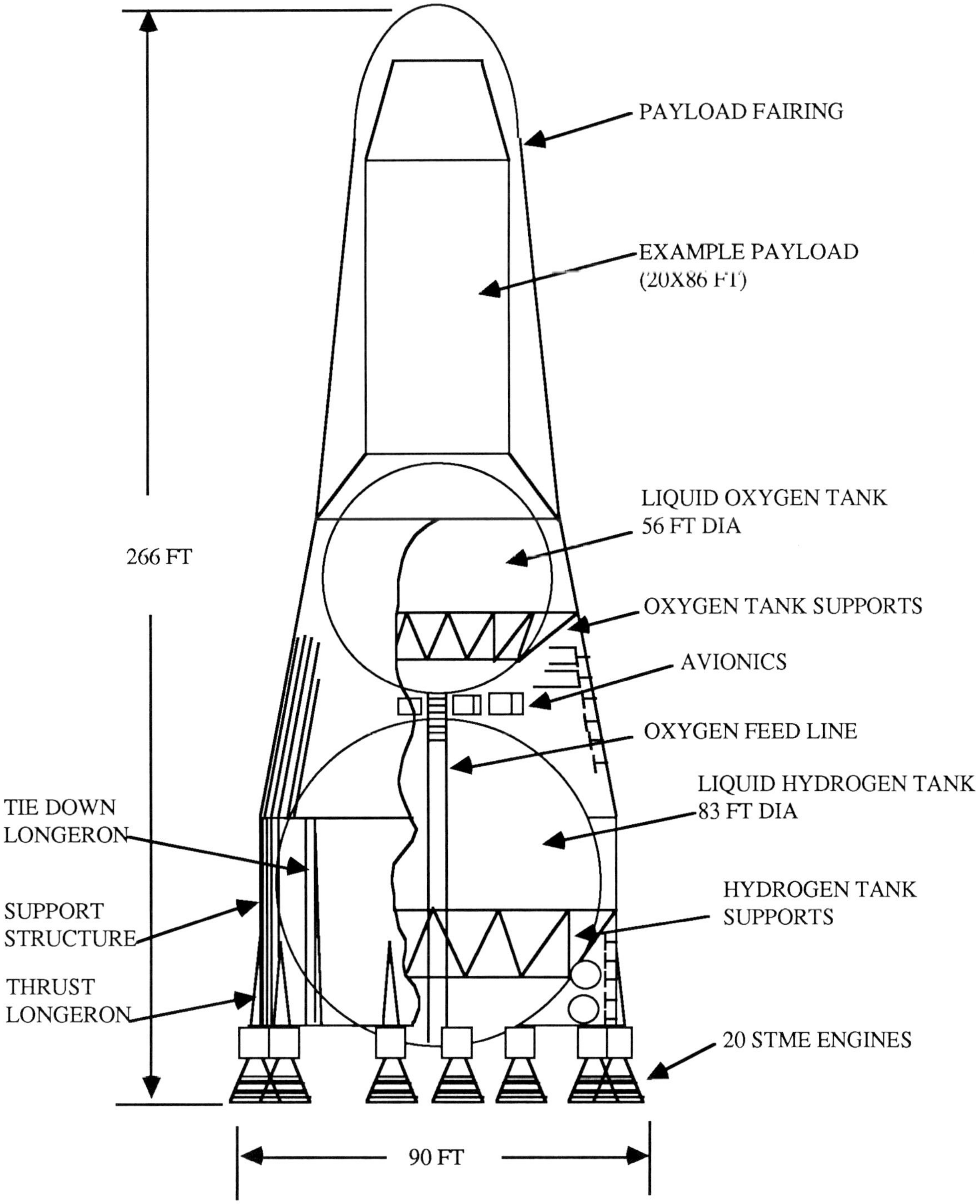

Perigee West Company's Euclid Single-Stage-to Orbit ULLV

General Dynamics Production facility for 120 foot diameter aluminum spheres in Charleston South Carolina

Exterior view of production facility showing sphere loaded on a transport barge.

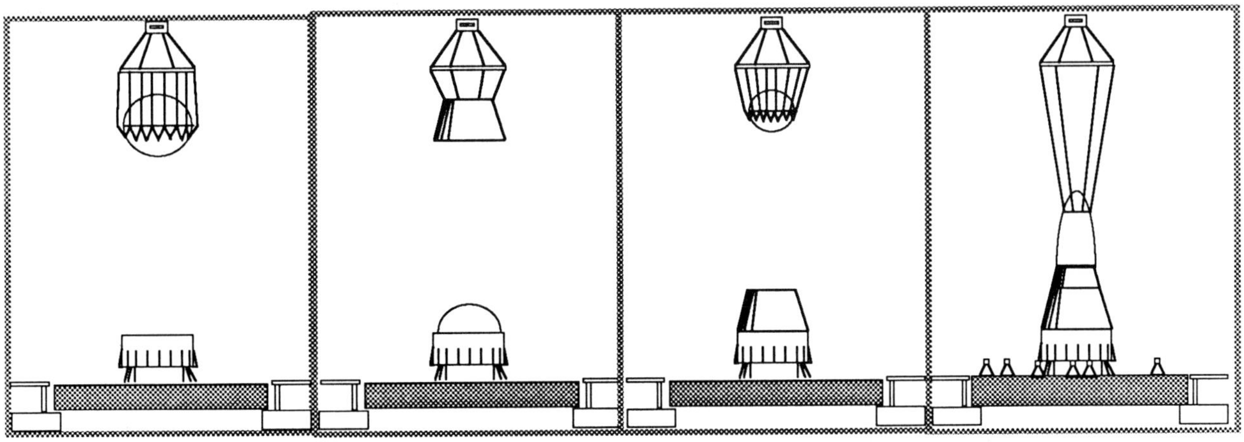

Multi-bay erection facility for Euclid ULLV. Shaded areas represent launch barges.

The launch barges are fitted internally with propellant loading equipment which can be remotely operated. One operations plan is to launch the ULLV on a sea platform. In this case the barge would dock at a loading station to take on propellants before being towed to the launch station. Hydrogen and oxygen could be generated locally, or they could be transported to local storage tanks from a distant production facility.

A second plan centers on land-based launch operations. In this arrangement the barges do not carry propellants. The vehicle is instead loaded from large storage tanks adjacent to the launch station. As in the case above, the propellants can be either manufactured locally or trucked in. Probably the major issues to be resolved in installations of this type are local environmental impacts. In a high launch rate situation, acoustic loads alone could be the driver for offshore operations.

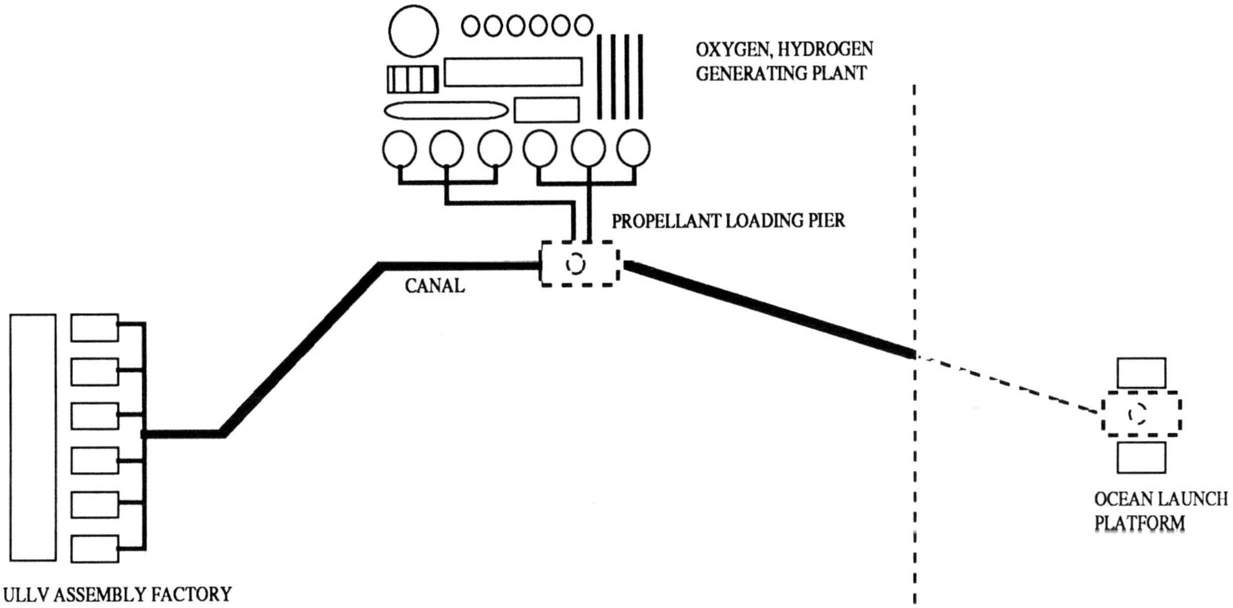

Topology of a ULLV launch operations concept based on offshore launches.

In the ocean launch operations scheme, the launch barge with the empty ULLV, secured to its launch pedestals, is towed to a location between twin ocean platforms. It is then ballasted to rest on a submerged section of the platform where it is secured in place. The launch vehicle is loaded by remote control from the propellant tanks inside the launch barge. With final checks complete, the vehicle is launched.

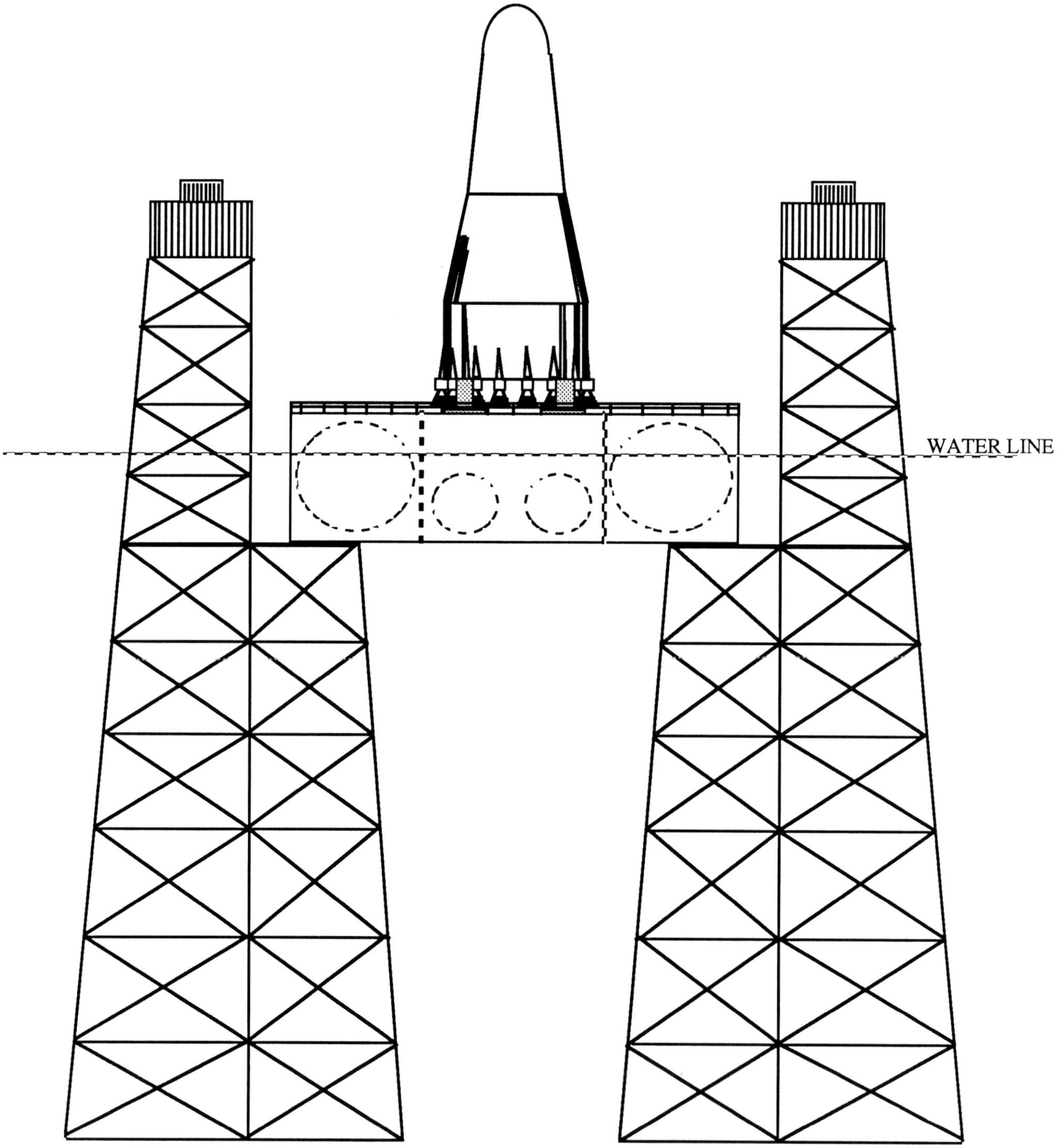

Euclid ULLV launch platform based on offshore launch operations

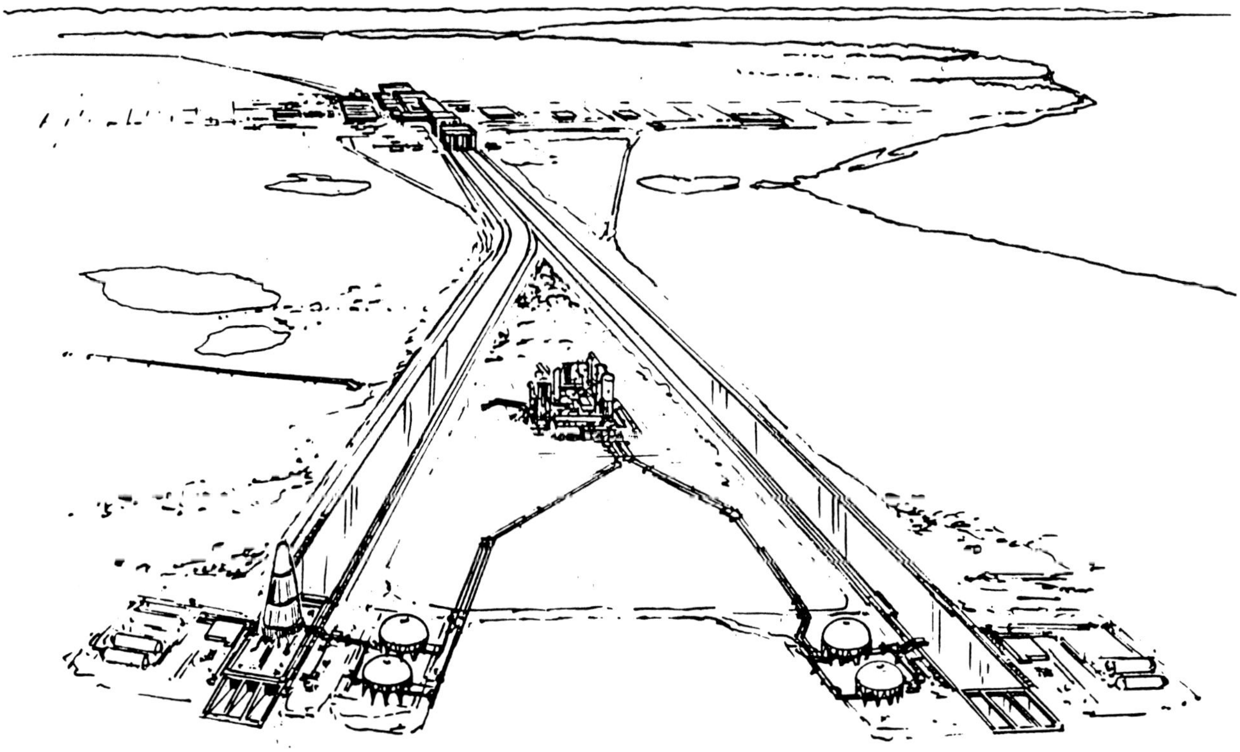

Land based ULLV launch operations concept showing canal access between the assembly building and the launch platforms.

While no missions, or even requirements currently exist for a ULLV, a future requirement could develop under certain scenarios.Two such scenarios are manned exploration of Mars and development of space power systems. The first will take eventually take place, perhaps within the lifetime of some of us. The second, along with fusion power, is one of two possible solutions to supplying the world's energy needs. In the not too distant future, one or both methods may be called upon to replace the non-renewable fossil fuel sources on which energy generation is presently largely dependent.

In the event that a manned Mars mission is undertaken, whether by the United States or as an international venture, the designers of the mission will be required to seriously address questions of crew safety, comfort, and health.

A necessary feature of the expedition equipment will be gravity spacecraft. It makes no sense to risk landing weakened or crippled astronauts on the surface of Mars, a condition that can result from many months in zero gravity. A second required feature will be safety. The spacecraft will have to contain sufficient redundancy, power and shielding to assure ability to get out of trouble, should problems occur during the mission. The third feature should be provisions for creature comfort. The Mars trip will be by far the most difficult manned mission yet tried. But the greater part of the mission will be time spent in transit. Most spacecraft functions will be automated to reduce possibilities of human error. ULLV's will make it possible to achieve the above capability. The luxury of very large lift capability provides designers with the margins needed for a truly robust system.

One plan for providing in-transit artificial gravity as well as safety and comfort assumes a mission that is performed with two large redundant spacecraft, coupled and rotated in a manner to produce artificial gravity. A meaningful manned Mars mission with these features could be accomplished with three ULLV flights. Two flights would be dedicated to launching the two spacecraft into low-earth orbit, where they would be coupled together. A third flight would launch a nuclear powered transfer stage. This stage would have sufficient capacity to accelerate the spacecraft pair toward Mars, as well as power the much lighter pair on the earth return trajectory at the conclusion of the mission.

The gravity spacecraft will be made possible by development of very high tensile material for cable construction. Commercial carbon fibers are already available with tensile strengths exceeding one million pounds per square inch. Materials of much higher strength are under development in several laboratories. A leading candidate may be derived from complex carbon molecules, recently discovered, termed "fullerenes." The name honors Buckminister Fuller, as the molecules exhibit an architecture similar to Fuller's geodesic domes. The carbon clusters are also known as "buckyballs." A new carbon fiber material designated "buckytubes," has followed the discovery of buckyballs.

When the two spacecraft are mechanically docked, dual cable-and-reel subsystems in the nose sections are interlocked. An integral part of the subsystem is a transfer car, operating along the cables, that enables crew passage between the spacecraft.

The propulsion unit has three major components, a hydrogen tank , a control avionics, and a thruster. It provides propulsion for the entire mission. The hydrogen tank is fitted with propellant recovery equipment to prevent loss of hydrogen during the years-long mission. Development of a high temperature reactor will simplify the system, allowing use of ordinary water as the propulsion medium. The rocket engine, with its associated pumping and control equipment, will consist of one or more thermonuclear reactors. Multiple reactors will be required in the event that an "engine out" capability is deemed necessary.

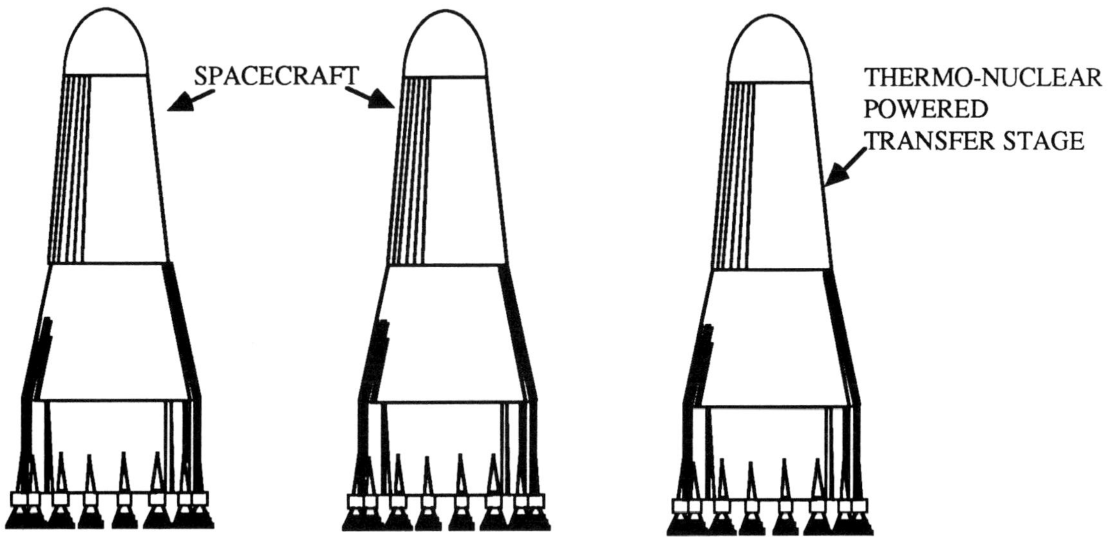

Three ULLV's, orbiting 300 tons each, will support a major manned Mars mission.

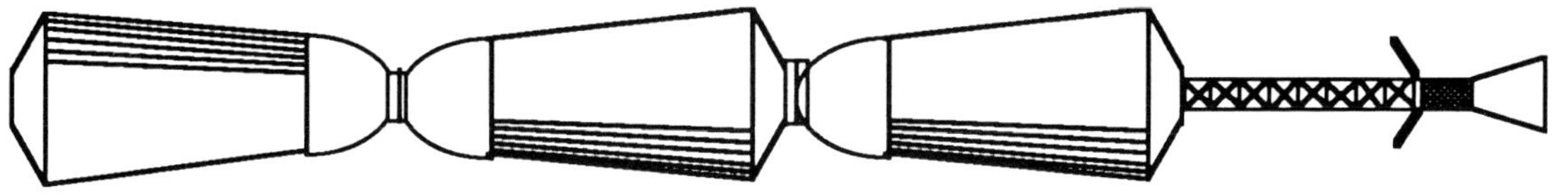

Once in Earth orbit, the twin spacecraft, containing landing and ascent stages, as well as habitats, are docked nose-to nose. A nuclear powered transfer stage is then docked to one of the spacecraft. It separates when Mars transfer velocity is reached.

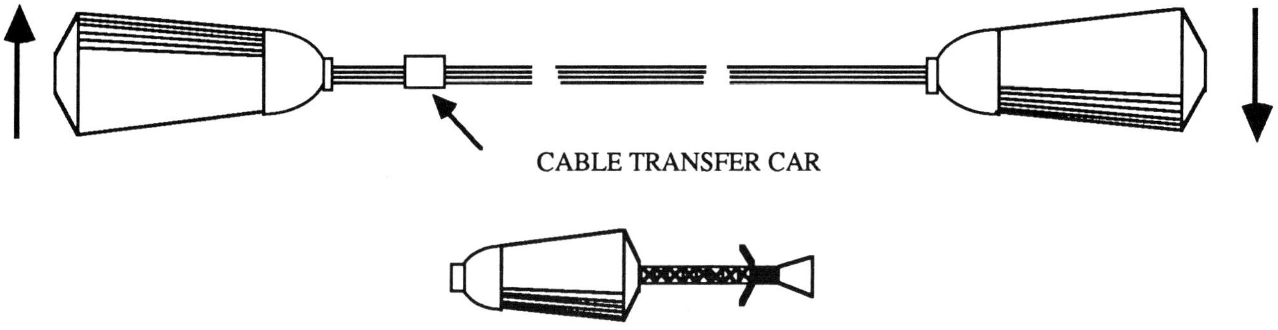

The spacecraft separate but remain attached by a reeled cable system. At the same time the couplet is placed into rotation. The spacecraft and transfer stage proceed separately to Mars proximity, where they will reconnect for Mars capture maneuvers..

12-5 Outlook

Except for isolated mental forays into space exploration and exploitation scenarios where ULLV's play a big role, little of significance is stirring mission planners in the direction of large-scale space exploration and exploitation efforts. The Space Exploration Initiative, announced by President Bush, died before any substantive work could be done. Even so, those who did engage in preliminary plans confined their thinking to launch capabilities offered by the National Launch System, or the Russian Energia; at most a payload lift capability of 250,000 pounds.

What could change the picture? In one scenario, NASA has the surface of Mars crawling with miniature robots, sending by radio signal so much data that most of it has to be stored for future examination. Highly automated terrain evaluation techniques developed for the U.S. army are used to perform the initial scans. The surveys produce incontestable evidence that a life form did live and flourish on Mars. Such a turn of events would be enough to trigger a cooperative international effort to organize and execute manned expeditions.

A second scenario encompasses a concept under which the entire in-space future transportation system would operate. This plan would use ULLV's to carry water into space, at a cost around $200 per pound. The water would be used as the propulsion medium in space based nuclear powered steam rockets. Water is dense, safe, and easy to store and pump. It is the ideal propellant for space vehicles. The economics of a steam rocket system can be shown to be far superior to any system conceived to date.

A third scenario converges on space-power generation. A growing awareness of the need to husband fossil fuels, which are being depleted at an alarming rate, can lead to more serious search for alternative energy sources sooner than most people expect. The principal alternatives are fusion power and solar energy. Fusion power is not expected to be developed for another fifty years. Substantial research remains to be done. But space-power systems are more in the realm of engineering. A pilot system to prove the feasibility of large scale space-power generation systems could be undertaken at any time. A pilot operation would be no small effort of itself, and would require ULLV's for accomplishment.

Solve the Earth's overpopulation problem by moving people off the planet? This futuristic idea, entertained by some, is however unrealistic. One could not build rockets fast enough to keep up with population growth. Nature will implement its own ways of handling overpopulation. Go to space as a means of preserving the species? Some would believe the idea quite valid, an ingredient of nature's strategic survival plan for humankind.

13

Nuclear Powered Rockets

13-1 Why Nuclear Power?

Nuclear powered rockets have been investigated since the early 1950's as a way of obtaining better performance than that afforded by chemical combustion engines. Performance is measured in terms of specific impulse, which is the rocket's thrust divided by the mass flow of propellant per second. Practical limits of chemical propulsion are reached with an oxygen/hydrogen reaction, which incidentally produces an exhaust product (steam) with the least environmental impact. Oxygen/hydrogen engines operating in space can yield a specific impulse of about 480, although engines produced and operated to date peak out at about 452. Some exotic propellant combinations yield better performance but are generally considered to be too hazardous for routine operations.

Two factors can improve specific impulse: raising the combustion chamber temperature; and lowering the molecular weight of the exhaust gases. Right off, hydrogen appears as the ideal candidate fuel at a molecular weight of 2, with oxygen (molecular weight 32) as the oxidizer. The temperature is determined by combustion kinetics at the mixture ratios chosen for thrust chamber steady -state operation. Exhaust temperatures of chemical rocket engines may be 2500 to 3600 degrees K. The chambers and exhaust nozzles must be protected from melting by internal cooling or by ablative insulation.

A higher specific impulse, about double that of chemical rocket engines, can be obtained with a nuclear reactor. In a propulsion application, hydrogen is used as the working fluid so the rocket exhaust has a molecular weight about one-eighth that of an oxygen/hydrogen engine. Having selected the optimum propellant, the remaining way to improve performance rests with operating at as high a temperature as possible. This is determined by the temperature at which the reactor can operate for long durations without melting or distorting its internal components. In a nuclear reactor, hydrogen propellant is heated to temperatures between 1800 and 2500 deg. K., and is then expanded and exhausted through the nozzle.

Nuclear powered rockets have been ruled out for operation in the Earth's atmosphere and therefore cannot be used for launch vehicles. But an application does exist for space-based transfer vehicles, operating, say, from Earth orbit to Lunar bases or between Earth and Mars. Due to the heavy weight of the engines, they do not yield expected performance gains when used on small vehicles, but on large vehicles, where they have a lesser affect on propellant mass fraction, they can nearly double the payload that can be carried by an equivalent chemical rocket.

Many different designs for nuclear rockets have been created over the past thirty-seven years. Scientists and engineers have been searching for structural arrangements that permit operation at higher temperatures, yielding even better performance than the factor-of-two promised by solid core reactors. Some concepts show potential for specific impulse as high as 1800, four times the capability of chemical propulsion rockets.

By far the greatest amount of work has been accomplished on solid core nuclear rockets, including substantial development work on several sizes of reactors. A typical solid core reactor operates in conjunction with a turbopump, which feeds hydrogen to the reactor. The liquid hydrogen is used to cool the walls of the chamber before entering the reactor. A portion of the reactor gas is bled off to operate the pump's turbine. In a rocket application, the effluent from the turbine can be ported into nozzles to provide thrust for roll control.

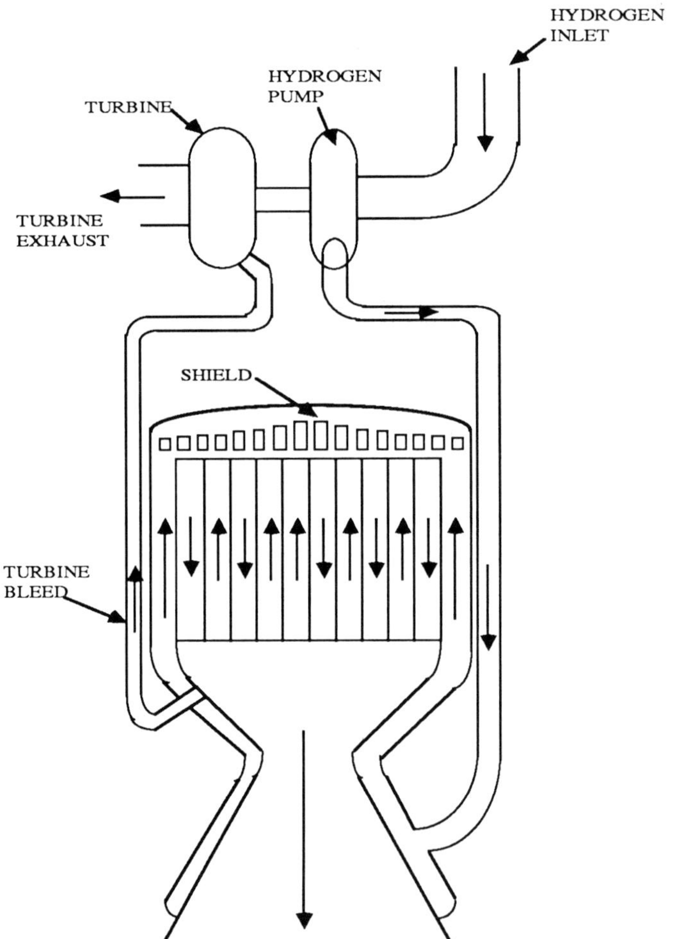

Schematic diagram of a solid core reactor with a "hot-bleed," amounting to about 3 % of total hydrogen flow, to operate the turbine drive.

Artist's concept of a nuclear powered rocket on a space mission (NASA)

The hydrogen-cooled, solid core reactor was selected for development in 1955. One of the out-year program objectives was to provide propulsion for a manned expedition to Mars. During the 1960's, nuclear rocketry played a significant role in the planning of interplanetary expeditions since it provided a means of significantly reducing transit time. Fast transit times reduce the transport masses required for the mission as a result of reduction of time-dependent consumables.

The U.S. nuclear rocket program continued until 1973 and produced operating reactors. But by this time U.S. space priorities had changed, resulting in termination of the program. Research and development work was initiated again in the late 1980's in a joint effort by NASA and the Energy Department. Once again, the effort was terminated in 1992 due to budget pressures and absence of firm requirements.

The more recent effort was leading to development of a hydrogen-cooled particle bed engine, which held promise of a specific impulse of about 1000, as well as much lighter weight than the solid core reactors. Atlas and Titan launch vehicle manufacturers were able to show substantial improvements in payload capability for near-Earth missions, using upper stages powered by the new reactor.

The 1955 initiative began the Rover program, the name given to the overall United States nuclear rocket development effort. The program began with the Kiwi series of reactors, which were the first to operate with liquid hydrogen. The final Kiwi model operated at 1890 degrees K for 11.3 minutes.

Following the Kiwi rcactors, the NRX series of developmental reactors was built. The goal was to demonstrate a specific impulse of 760 for one hour in 1100 MW reactor. The thrust level was 55,000 pounds. The goals were exceeded in the NRX-6 test, which ran for 62 minutes at 2,220 degrees K.

The Phoebus series of research reactors followed the Kiwi program. The goals of this programwere to demonstrate a specific impulse of 825 and to conduct experiments in adjusting power density and power level. These goals were demonstrated in the Phoebus-1B and Phoebus -2A tests. The Phoebus -2A test operated for 12 minutes.

Two other reactor series, Peewee and Nuclear Furnace (NF), were designed to demonstrate high temperature and long-life fuel elements. Peewee-1 ran at 2,555 degrees K and 514 MW for 40 minutes. NF-1 ran at 2450 degrees K and 54 MW for 109 minutes.

In September of 1964, the first tests were performed on the NRX-A2, a reactor destined for rocket propulsion application. NERVA, an acronym for "nuclear engine for rocket vehicle application," reached full power of 1100 MW and was operated for approximately five minutes. In March of 1966 NRX/EST, the rocket engine breadboard power plant operated at full power for 13.5 minutes. The fifth test in the NERVA engine series exceeded the design goal of 60 minutes at 1100. The most powerful nuclear rocket reactor was Phoebus-2A, which was tested in June of 1968 for 12 minutes above 4000 MW.

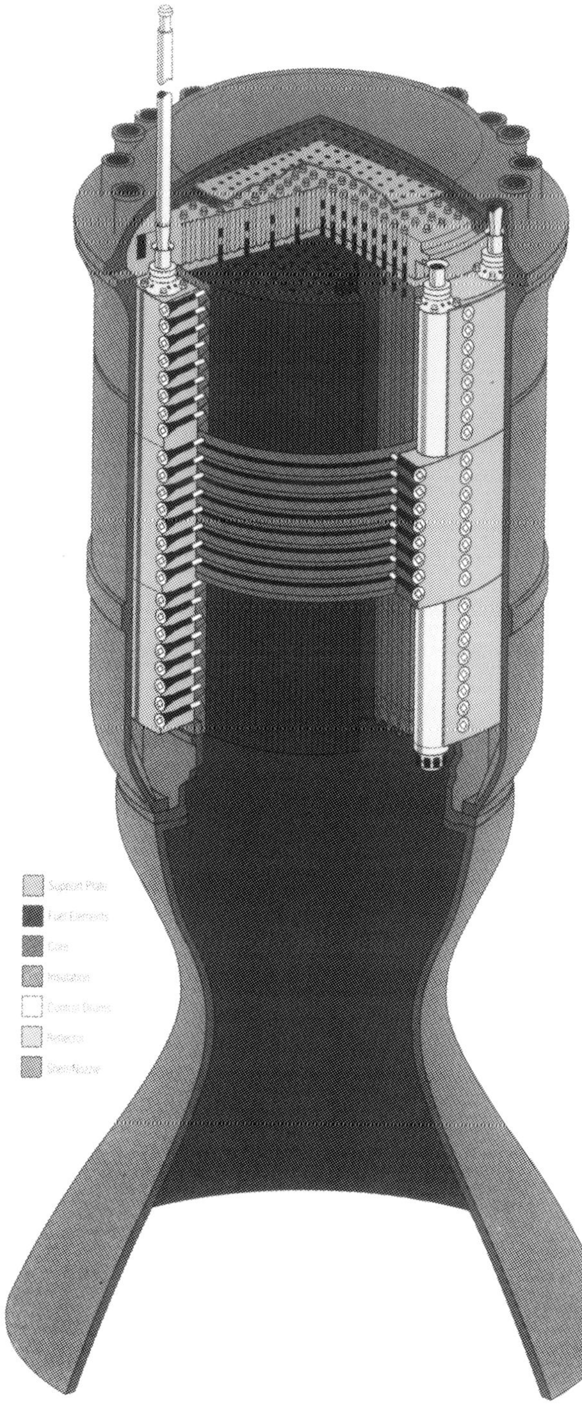

A cutaway view of the Rover nuclear rocket reactor (Rocketdyne)

A 50,000 pound thrust flight configuration nuclear reactor engine installed in a Nevada test stand.

Phoebus 2A nuclear thruster being transported to a Nevada desert test cell.

The NERVA program also produced what was called the XE' "small engine." It was designed for a thrust of 16,000 pounds with a specific impulse goal of 825. The engine had a total mass of 5620 pounds, about seventeen times the weight of the Pratt & Whitney RL-10 chemical rocket engine, which powers the Centaur upper stage. This engine provides an interesting perspective regarding applicability for small stages since it weighs more than the entire Centaur upper stage.

The XE' represented an accumulation of all of the knowledge that had been gained in the nuclear rocket program. It used hydrogen as the working fluid and utilized a topping cycle turbine drive, so that propellant used to drive the turbine was fed back into the reactor rather than being mostly wasted. A single stage turbine and single stage pump sufficed for the feed system. The chamber was regeneratively cooled with the exception of a short section of the nozzle skirt. A radiation shield of borated zirconium hydride was installed above the reactor to reduce neutron heating of the hydrogen propellant tank situated ahead of the engine.

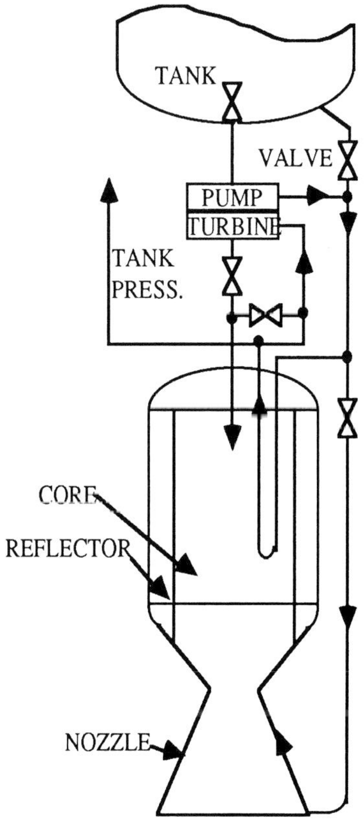

HYDROGEN PROPELLANT

FULL FLOW TOPPING CYCLE

SINGLE STAGE PUMP

SINGLE STAGE TURBINE

BORATED ZIRCONIUM HYDRIDE
RADIATION SHIELD

6 CONTROL DRUM ACTUATORS

5 VALVES

REGEN. COOLED NOZZLE

OVERALL LENGTH -174 INCHES

TOTAL MASS - 5620 POUNDS

Schematic diagram and description of the 16,000 pound thrust "Small Engine nuclear rocket design.

McDonnell Douglas prepared preliminary designs for a nuclear powered stage that could be orbited by the Space Shuttle. The stage weighed 51,000 pounds fully loaded and required the entire cargo bay for the stage alone, with no payload. At that, it was necessary to hinge a section of the engine's nozzle sideways to effect a fit. But the same safety concerns that caused NASA to cancel usage of the hydrogen powered Centaur stage with the Space Shuttle applied to the hydrogen fueled nuclear stage. NASA was not ready to carry tanks of hydrogen in the Space Shuttle's cargo bay.

The Rover and NERVA programs, conducted over a fifteen year period, were remarkable technical accomplishments. One after another, the technical goals set for the various development efforts were met and exceeded. The result was a propulsion capability that offers opportunity to double the performance of space transfer vehicles.

Environmental issues limit the application to launch vehicle upper stages and space-based transfer vehicles. In all probability, nuclear engine costs will rule out use of nuclear reactors on expendable stages, except for certain deep-space missions. Their future is more likely in scenarios that demand economical flight between the Earth and the moon, and between the Earth and Mars. The logistics system that will be needed to support such scenarios will include orbiting service stations for nuclear powered transfer vehicles, and hydrogen tankers operating from Earth for reloading the vehicles.

The recent resurgence of interest in nuclear rockets by NASA and the Department of Energy, looking to applications in the Space Exploration Initiative, shows that the subject is by no means closed.

13-3 Nuclear Powered Space Transfer Rocket

A nuclear powered rocket is simpler than a chemically propelled vehicle only in that it does not require two propellants, and hence an extra tank and pump. All other subsystems needed to make a rocket work are needed, including the main thruster, avionics, control thrusters, structure and propellant tank. The propellant tank will require thermal insulation of a type that minimizes hydrogen boiloff. "Super insulation," comprised of many layers of reflective plastic film, separated by screen-like fabric, has already been developed for this purpose. For missions that last months or years, a hydrogen subcooler is required to prevent any loss of propellant.

In a nuclear rocket, the hydrogen pump may be contained inside the tank, enabling it to operate at very low inlet pressures. Pressurization, if needed, is accomplished by a hydrogen bleed from the turbopump which is vaporized and warmed by heat from the engine.

An interesting aspect of a nuclear rocket is that hydrogen is the only fluid needed. Cold hydrogen gas is an acceptable medium for the vehicle's small control rockets and is available from the hydrogen tank.

Compared to conventional chemically propelled stages, nuclear rockets will not be cheap. Their main advantages lie in the prospect of doubling the performance of chemical rockets as well as a service life that could be hundreds of hours. This leads directly to the conclusion that such systems should be designed from the beginning to be reusable. As such, the supporting space infrastructure must be designed and installed concurrently with activation of a nuclear transport capability. This will involve creation of orbital tanking and stage servicing depots as well as an earth-to-orbit propellant transport system for replenishing the stored depot propellants.

As a reusable system, the nuclear rocket will be fitted with a docking mechanism at its forward end. This mechanism will enable union with all manner of spacecraft fitted with similar equipment. It will serve also as the junction between the craft and fuel depots for reloading propellant, as well as docking to other stations for servicing. In its initial flight to earth orbit aboard a heavy lift vehicle, the nuclear rocket could be loaded with fuel for its initial missions. After that, all servicing must be done from orbital facilities or special tankers sent up from Earth.

A significant aspect of nuclear rocket design strategy will be a determination of the capacity of the hydrogen tank, through design and operations trade-offs. A reusable nuclear rocket may turn out to operate most economically in a role analogous to Earth-bound semi's, which often transport partial loads. In such a circumstance, the vehicle would load up with only the hydrogen needed to accomplish a specific mission.

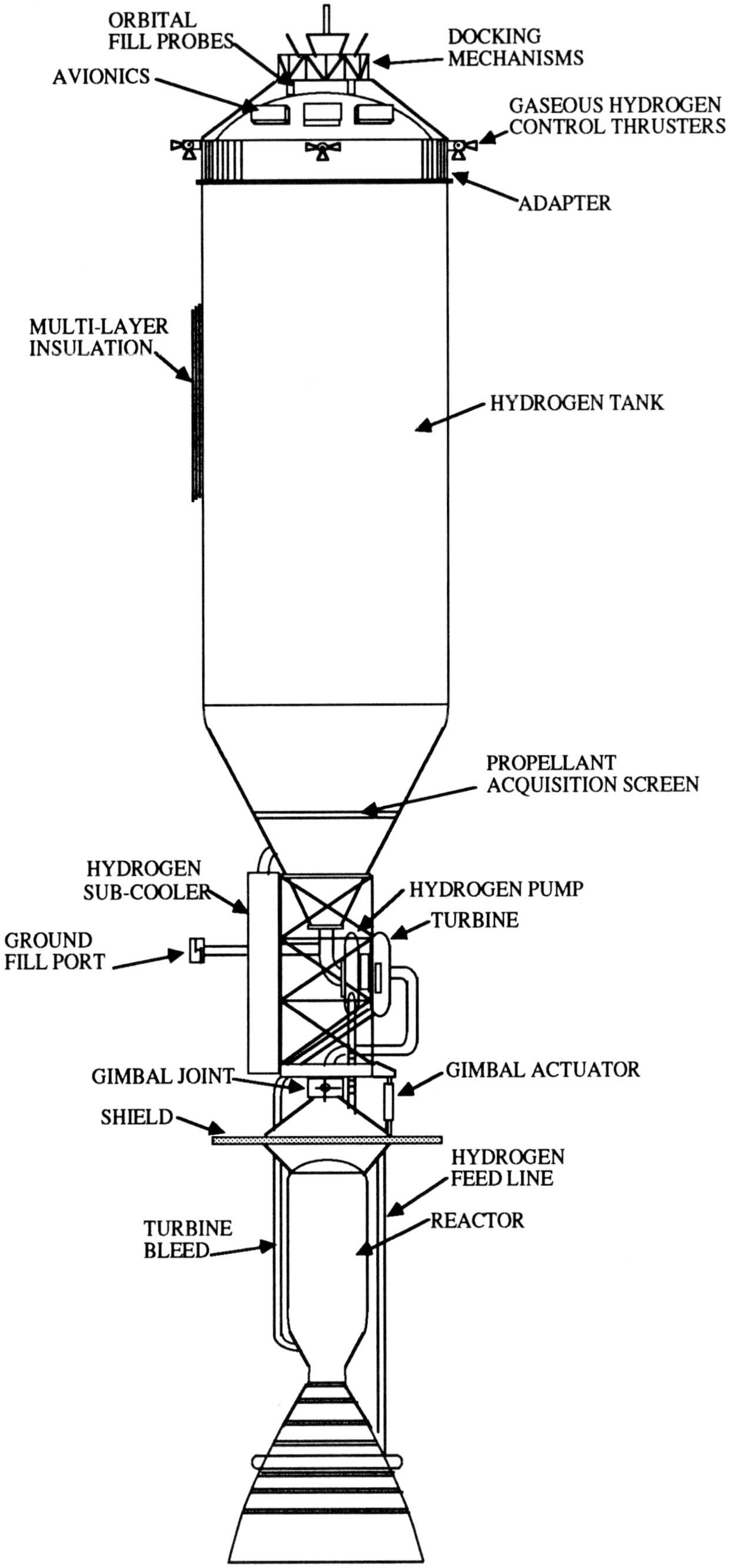

A reusable nuclear powered stage that might be used for lunar transport or shuttling payloads to and from geosynchronous orbit.

13-4 Outlook

It sometimes happens that a particular technology will leap far in advance of any immediate requirement, simmering on the sidelines for years, waiting for a mission. Such is the case for nuclear powered rockets. U.S and Russian engineers have both achieved workable propulsion systems in an assortment of sizes, and neither nation is close to an application. When it does happen, granting that the spirit of international cooperation in space ventures proceeds as many predict it will, it is likely that the future nuclear rocket power plant will be a product of the technologies of both nations.

Today, a development program for a nuclear rocket engine must face problems that were ignored during previous programs, adding considerable expense to the effort. Environmental constraints will require that all exhaust products from test firings be captured and processed. Previous testing allowed radioactive exhaust to dissipate in the atmosphere. The cost of installation and operation of a waste collection facility, as well as the cost of disposal, will constitute large added costs to development and production programs.

Planners of future programs would be well advised to treat nuclear rockets as one part of a larger, interrelated system. Nuclear rockets will be sterling performers, but to be economical they must be reusable. Reusability implies a major space-based and land-based infrastructure to provide vehicle servicing and propellant supply. Such considerations as what to do with a worn-out nuclear powered vehicle are a valid and essential part of a total systems study. Enough rocket studies have been performed to confirm feasibility and benefits of nuclear propulsion. System studies are now needed to identify, trade and design all of the elements that will be needed for routine space transportation operations.

Continuing studies and development work may produce a nuclear reactor more advanced in concept than the solid core reactors that were tested in the Rover program. Objectives of engineers who work in this area are to produce a much lighter weight engine, as well as improve performance by raising the temperature at which the device can operate. More about this in Chapter 15, "Advanced Propulsion Concepts."

The Department of Defense, NASA, and the Department of Energy continue to fund research and development efforts in nuclear propulsion, but under severely constrained funding. Congressional support varies from year to year, but at present it is nearly nonexistent and future funding is in question. In the absence of a requirement, much less a mission, nuclear rocketry must be considered an area of space technology that at best merits only modest expenditures. It is a technology in search of a requirement, about which one can now only speculate.

14

Chemical Propulsion Orbit Transfer Stages

14-1 The Nature of the Business

If booster rockets are the Clydesdales of the rocket industry, then orbit transfer vehicles are surely thoroughbred racers. The difference is that a boost stage can afford to have extra poundage in its structure and other hardware, whereas any extra weight in a transfer stage subtracts from its payload carrying capability. The reason, of course, is that a booster carries a large weight above it, consisting of other stages and the payload, with the payload being but a small part of the whole.

Orbit transfer stages are where rocket designers stretch their expertise, to extend the capabilities of materials of construction as far as possible. The greatest payoff comes with the lightest rocket, regardless of the method of propulsion employed.

Chemical upper stages cover a broad spectrum of capabilities and are manufactured by several companies. Most are solid propellant rockets. The only U.S. operating liquid propellant upper stages are Centaur, Delta's SSPS, and specialized devices used in DoD classified programs. Centaur has already been mentioned in Chapter 2, as the upper stage of Atlas, and in Chapter 3, as the upper stage of Titan IV.

Solid propellant stages are of two general types: stages that are used with spacecraft that are spin stabilized for powered flight; and stages that do not spin, but are instead provided with three axis control. In the first type, the stage consists of nothing more than a solid rocket motor containing a propellant charge that is sized for the energy required to accelerate a particular payload. In the second case, the stage is equipped with all guidance and control functions necessary to operate in a three-axis stabilized mode. The rocket motor itself may be equipped with a swivel nozzle for steering. An attitude control system, consisting of small liquid fueled rockets, is required for roll control and other functions.

In another class of three-axis stabilized stages, guidance and control functions are provided by the spacecraft. In this case, the propulsion device is again a simple rocket motor, like those used in spin stabilized spacecraft.

Principal manufacturers of solid rocket motor upper stages are Thiokol Space Operations of Brigham City, Utah; United Technologies' Chemical Systems Division of San Jose, California; and Hercules Aerospace Company, of Wilmington, Delaware. Integration of the motors with spacecraft, or conversion to a higher capability, such as three-axis stabilization, is usually accomplished by systems contractors like

McDonnell Douglas, Martin Marietta, or Boeing Defense and Space Group.

Historically, liquid propellant orbit transfer stages are few in number. Lockheed's Agena, which flew atop Atlas and Titan in the 1960's, was the first to be applied to numerous DoD and NASA missions. Early NASA missions participated in the manned Gemini flights that followed the Mercury program, and the Ranger series of Lunar exploration missions. Agena was used for numerous DoD flights which remain classified.

McDonnell Douglas' S-IVB was used only in the Apollo program. It provided orbit transfer propulsion for the Lunar missions, which consisted of the Lunar Landing Module, the Service Module, and the Apollo Command module, weighing together approximately 50 tons. Similar to Centaur, the SIVB was a high energy stage, using liquid hydrogen for fuel. It used a much larger engine, Rocketdyne's J-2, which developed a thrust of 250,000 pounds.

Apollo also carried a second liquid propellant transfer stage, the Apollo Service Module, which provided propulsion for the Command Module's return trip to Earth.

Centaur, the high energy hydrogen fueled stage developed and built by General Dynamics, has flown more missions than any other liquid-fueled transfer stage. It has seen service on Atlas and on Titan II. This vehicle has an illustrious history of accomplishment, highlighted by such missions as Surveyor, Pioneer, Voyager, Mariner and Viking. Aside from the NASA missions, Centaur has also been used extensively for commercial and military missions.

Centaur began its development as a DoD program in the late 1950's. As part of this initiative Pratt & Whitney was contracted to develop the first hydrogen/oxygen rocket engine. As the Air Force's attention turned to other space projects, including Dyna-Soar, which did not need Centaur, the program was given over to NASA for continuation. The project eventually ended up under the control of Lewis Laboratories in Cleveland, Ohio, under whose direction General Dynamics completed development of the stage.

Centaur and Delta's SSPS remain the only U.S. liquid propellant transfer stages in production. In most Delta missions, the SSPS does not function as a transfer stage, however. Centaur comes close to delivering the best performance that can be achieved with a chemical propulsion device. Use of thin-wall stainless steel tanks has enabled designers to produce a comparatively lightweight vehicle. Even so, some improvement can still be obtained by installing lightweight avionics and a high performance single engine.

United Technologies Chemical Systems Division produces an array of space motors under the Trade name of Orbus, for "Orbital Bus." These motors are used for spacecraft orbit transfer and as upper stages for launch vehicles. The family includes the Orbus 21 and 6E, produced for the U.S. Air Force/Boeing Inertial Upper Stage (IUS) and used as an orbit transfer stage on Titan IV and Space Shuttle.

Orbus 21S is a spin-stabilized variant of Orbus 21. It was produced for launching Intelsat VI communications satellites, placing the spacecraft into a geosynchronous transfer orbit. This motor utilizes a Kevlar/epoxy case and a carbon-carbon exit cone. Its top propellant load is 21,200 pounds, but it can be fired with propellant loads as low as 10,700 pounds.

The Orbus 6E is the largest U.S. space apogee motor, used to circularize spacecraft orbits at geosynchronous altitude. The motor has a segmented cone which can be extended in flight to gain an additional 5% in specific impulse. The motor is 6.5 feet long when compacted and 10 feet long with the nozzle extended. It contains 6,000 pounds of propellant and generates an average thrust of 18,200 pounds.

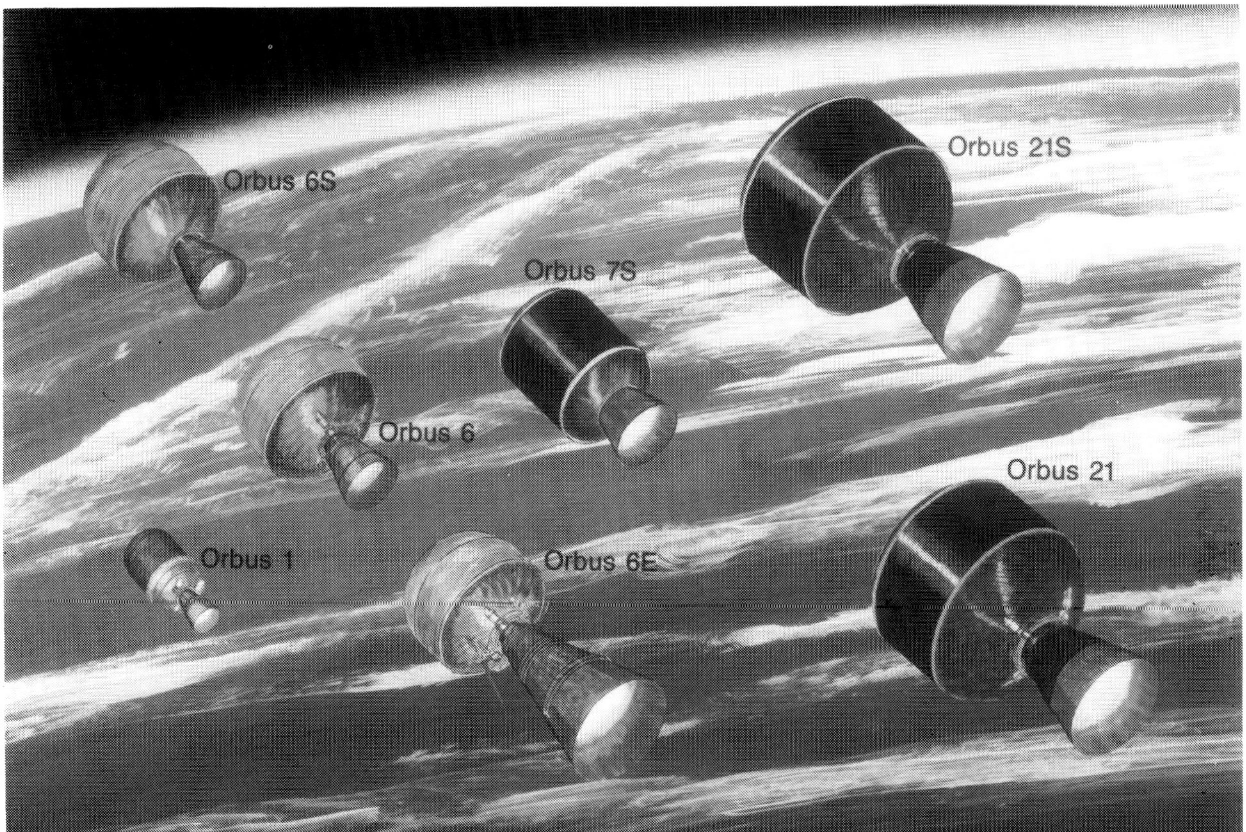

The Orbus family of of solid propellant space motors (Chemical Systems Division)

The Orbus 6S is a spin-stabilized motor produced for NASA IUS planetary missions. It is identical to the 6E except that thrust vector controls used on the 6E are removed. The Orbus 7E is produced as a spin-stabilized perigee motor for the launch of Hughes HS-393 series communications satellites. It has a 7300 pound load of propellant.

Chemical Systems Division's Orbus 21 motor provides primary propulsion for Orbital Science's Transfer Orbit Stage (TOS). TOS is a single-stage solid propellant vehicle, designed to work as a transfer stage in conjunction with either the Space Shuttle or Titan. The stage weighs approximately 24,000 pounds when ready for launch and measures 11.5 feet in diameter by 13 feet long. This vehicle launched the Mars Observer spacecraft in December of 1992 atop a Titan III.

Thiokol manufactures the STAR family of high performance space motors for spacecraft and as upper stages for launch vehicles. The motors cover a broad range of dimensions and weight. Diameters range from five to 75 inches and lengths from 8 to 102 inches. Propellant weights vary from 10 to 17,783 pounds. Illustrated is the STAR 37 motor that was used on seven Surveyor flights to the moon. This motor decelerated the spacecraft from 9000 feet per second to 400 feet per second. PAM motors are frequently used on communications satellites, providing the impulse needed to place them into a geosynchronous transfer orbit, following which a second solid motor provides the circularization impulse.

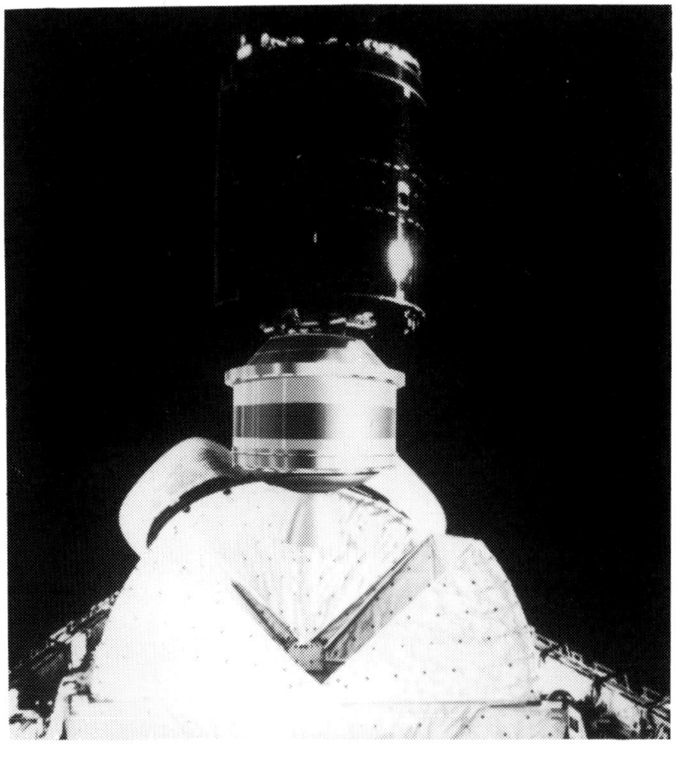

PAM D-2 solid motor shown while the Telesat-E Satellite is spin-released from the Space Shuttle Cargo Bay (Thiokol)

Boeing Defense and Space Group produces the Inertial Upper Stage (IUS), a two stage, three-axis stabilized vehicle with solid motor propulsion. It serves as an upper stage for Titan IV and the Space Shuttle. Missions for the 23 vehicles that have been built include spacecraft to the other planets and the sun, and placing satellites in geosynchronous orbit.

The IUS was developed and is built under contract with the U.S. Air Force Space Systems Division. The vehicle was intended initially to be a temporary stand-in for a liquid fueled "space tug" and was titled the "Interim Upper Stage." The word "inertial" replaced 'interim" in the title when prospects for building a space tug receded and it was realized that the IUS would see service for some time into the future. Development of the IUS was begun in 1976. The Space Systems Division provides the IUS to NASA for use with the Space Shuttle.

Launched by Titan IV or the Space Shuttle, IUS is capable of placing 5000 pound payloads into geosynchronous orbit or 8000 pounds on a planetary mission.

In a typical geosynchronous mission from the Space Shuttle the IUS is first rotated to an upward position in the orbiter's cargo bay where final checks of the satellite and IUS are made. The IUS with its payload is then ejected from the cargo bay. The IUS on-board computers then direct a series of preparatory maneuvers and fire the first stage motor for approximately 140 seconds to place IUS and payload on a geosynchronous transfer trajectory. The first stage is then separated from the spacecraft. About six hours later, after reaching geosynchronous altitude, the second stage motor is burned, placing the spacecraft in a circular orbit. The IUS is then separated from the satellite and moved to a non-interfering orbit.

The IUS vehicle is 17 feet long, 9.5 feet in diameter and weighs 32,500 pounds. It consists of an aft skirt, an aft-stage solid rocket motor containing 21,400 pounds of propellant and generating 45,600 pounds of thrust; an interstage adapter, a forward stage solid rocket motor with 6000 pounds of propellant, generating 18,000 pounds of thrust; and an equipment support section. The equipment support section contains the vehicle avionics subsystems. The solid motors for the IUS are built by United Technologies Chemical Systems Division.

When NASA cancelled the Centaur as an upper stage for the Space Shuttle, the Galileo science mission was left without a means of transportation. The decision caused a major redesign of the spacecraft and the flight trajectory to enable flight on the next best space transfer stage, the IUS. NASA has no high energy stage capability associated with the Space Shuttle. Missions requiring greater capability than that afforded by solid propulsion stages have to fly on Titan IV.

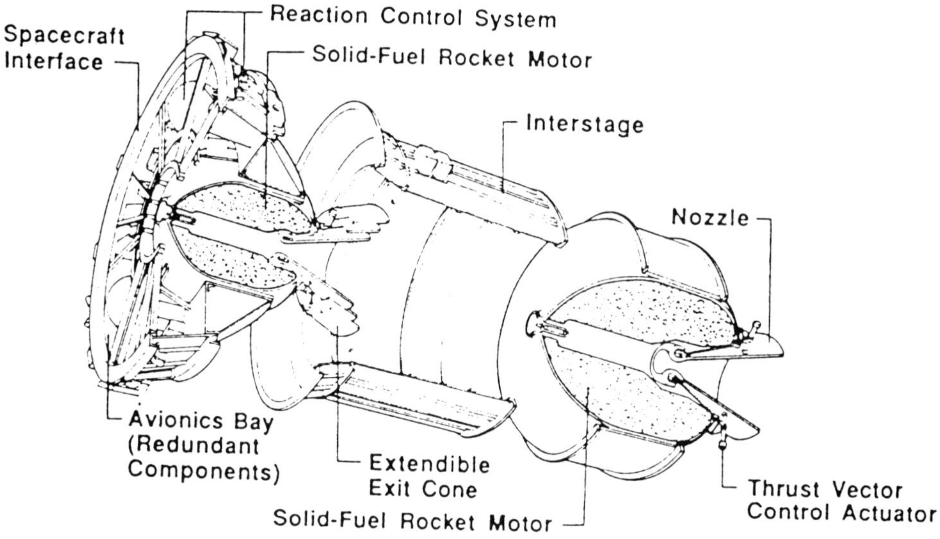

A breakdown of the Inertial Upper Stage (Boeing)

IUS with the Tracking and Data Relay Satellite being released from the Space Shuttle (NASA)

14-3 Hybrid Transfer Stages

A utilitarian upper stage that incorporates the best features of solid-fueled and liquid fueled stages can be obtained in a hybrid rocket, which would use a liquid oxidizer and a solid fuel for propellants. Such devices have not yet been built but are a relatively easy development. A significant advantage possessed by a hybrid stage is its ability to program multiple burns and to terminate firing when desired. A solid rocket motor cannot be shut off, necessitating the pre-loading of a propellant grain during manufacturing that is matched to the spacecraft mass that it has to accelerate. Multiple burns with a hybrid permit use of the most efficient trajectories to accomplish a given mission. From a ground handling standpoint, the hybrid is completely safe, eliminating the special handling procedures that are required for solid motors.

As compared to liquid propellant transfer stages, a hybrid stage has the same multiple burn features, but it is simpler in construction and requires fewer operations to prepare it for flight. This is because only one tank needs to be loaded, and no turbopumps are required. Unlike a liquid propellant stage, the hybrid has no potential as a reusable system, but it would be cheaper to manufacture as an expendable article. For future manned flight, for example as an upper stage for launching a personnel carrier, its inherent safety is a decided advantage.

The hybrid OTV is capable of longer orbit stay times than a hydrogen fueled vehicle without special thermal protection. Inhibiting liquid oxygen boiloff is a much smaller problem for liquid oxygen than liquid hydrogen due to its higher boiling point. If the oxidizer is nitrogen tetroxide, there is no boiloff problem.

A hybrid OTV can have a specific impulse approaching 340, which is respectably higher than that attainable with solid rocket motors, but about a hundred points lower than that attainable with a hydrogen fueled vehicle.

A hybrid transfer stage like the one illustrated would use a pressurized oxygen tank for feeding oxidizer to the fuel chamber instead of a turbopump. High pressure helium bottles are the source of pressurization gas. Loading of the oxygen tank is simplified by coupling the tank to the booster oxidizer tank during the loading operation. In the absence of any need to service the upper stage during ground operations, the launch pad can be designed without gantries.

Among potential manufacturers of hybrid motors for upper stages are Chemical Systems Division, American Rocket Company and Thiokol. Manufacture of a complete stage would fall to one of the existing systems contractors like McDonnell Douglas or Lockheed-Martin.

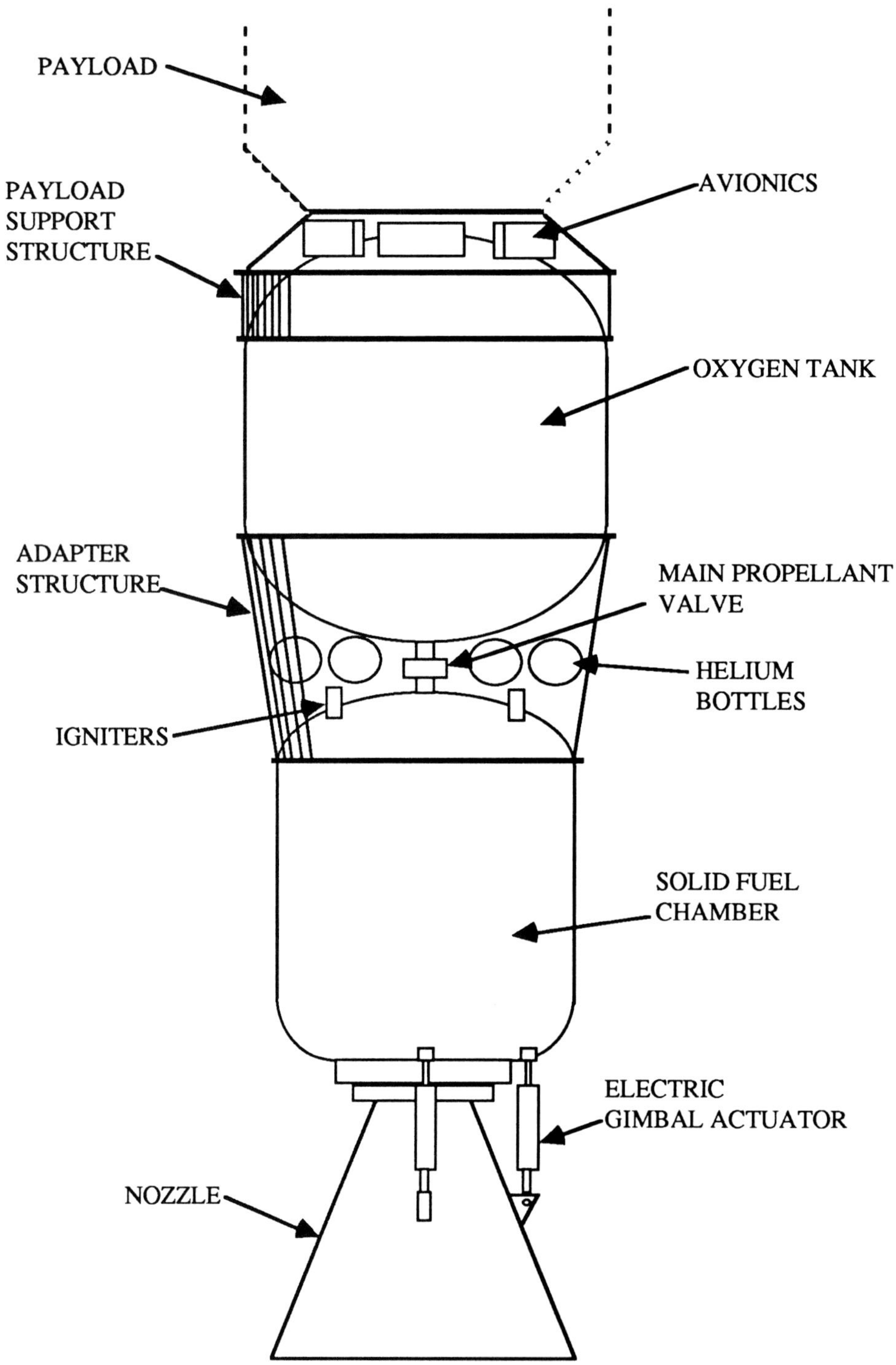

Main components of a hybrid upper stage or orbit transfer vehicle (Perigee West Company)

Over the past three decades, NASA and the Air Force have funded numerous studies of high energy space rockets under such titles as "Space Tug," "Advanced Maneuverable Propulsion System(AMPS)," On-Orbit System(OOS)," and "Orbit Transfer Vehicle (OTV)." Most were hydrogen fueled in the interests of maximizing performance, although some were based on "space storable" propellants like liquid oxygen and liquid methane in efforts to minimize propellant losses through boiloff. The AMPS utilized liquid fluorine for its oxidizer, which yielded better performance than oxygen in combination with liquid hydrogen.

The OTV's were planned for a hundred or more reuses, shuttling to mission sites and returning to an Earth orbiting base to be readied for another payload. Studies showed that considerable savings could be achieved over expendable transfer vehicles, once the supporting infrastructure was in place. The studies therefore included methods of servicing and refueling the rockets in space. The most recent ones concentrated on OTV's that would be based at Space Station Freedom. When the station's mission as a base for OTV's was dropped, the basis for continued work on OTV's largely disappeared.

An interesting aspect of the OTV studies was the employment of aerobraking as a means of saving fuel during a round trip mission. Instead of burning fuel to decelerate into low-earth orbit on the return trip, the OTV would skim the Earth's atmosphere, using air resistance to slow it down. To protect the craft from aero-heating, a large protective heat shield was deployed, which took the brunt of the heat load. The maneuver might be performed several times to achieve the desired apogee, at which time the craft's engines would be fired to place the OTV into a circular orbit. With this technique, large savings in propellant consumption could be shown, which essentially meant that the OTV could be a much smaller vehicle.

NASA also studied reusable OTV's that used the Space Shuttle as an operating base. In these schemes, the reusable OTV was Earth-based. Operational plans involved lifting the OTV with its payload into orbit and sending it on its mission. The OTV would then return, rendezvous with the orbiter, and be returned to Earth to be prepared for another run. In some cases the payload might be taken into orbit by a second Shuttle and joined to the OTV when in orbit.

As in the case of Shuttle/Centaur, NASA abandoned any ideas of carrying hydrogen fueled rockets in the Shuttle cargo bay after the Challenger accident. The decision was in fact made for the Agency when Congress mandated that the Shuttle's missions could no longer include transportation of commercial payloads.

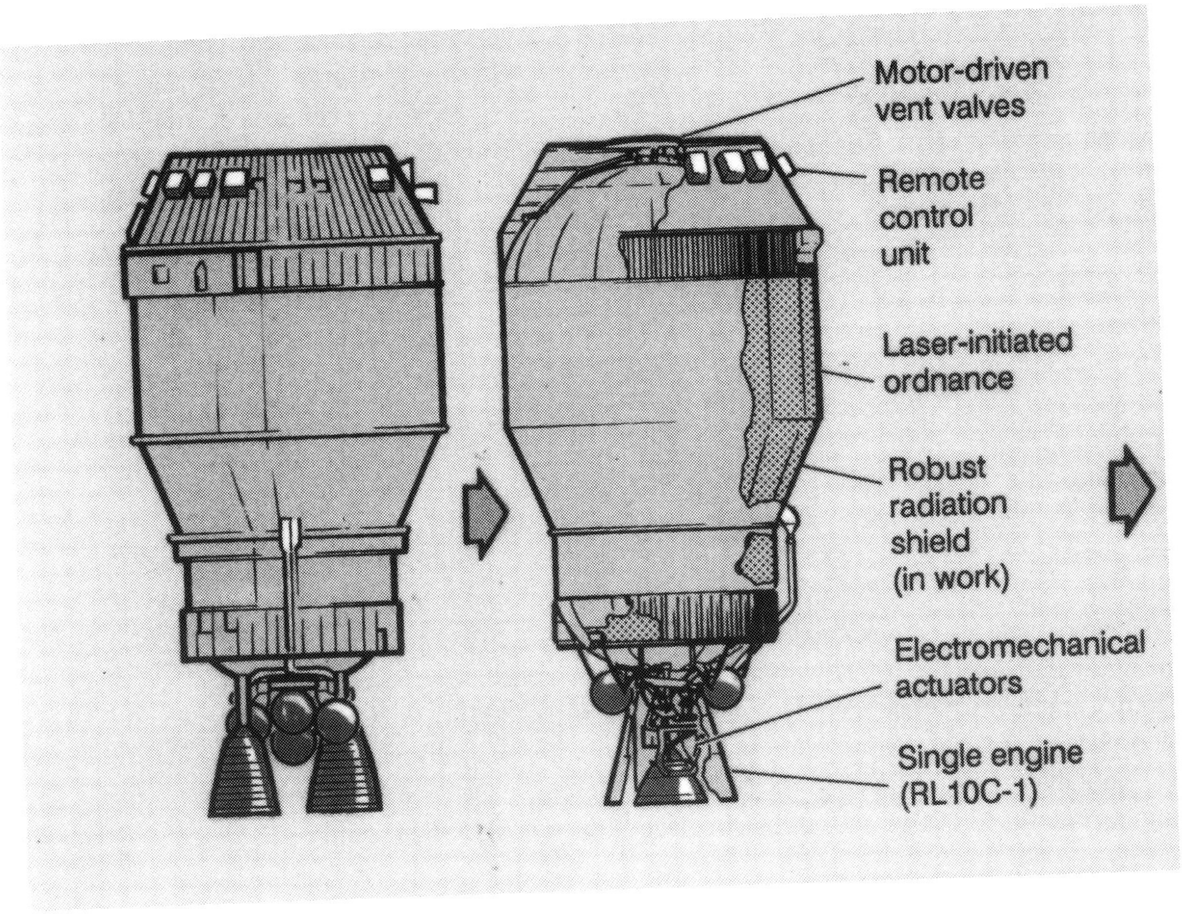

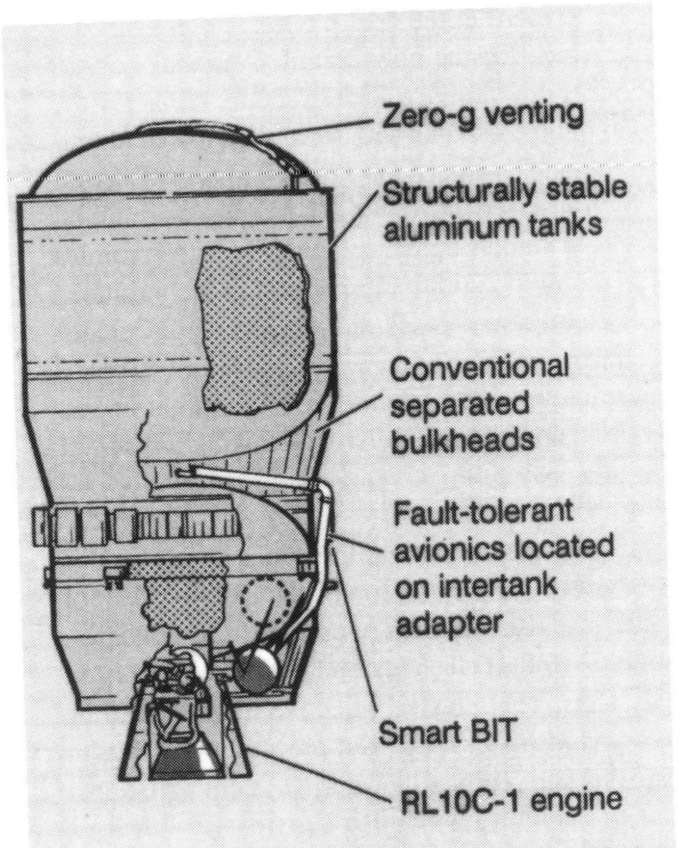

Centaur evolutionary path leading to high energy upper stage capability for the next generation launch vehicle (General Dynamics)

14-5 Orbit Maneuver Vehicles (OMV)

When NASA first undertook the Space Station program, one of the system requirements centered on availability of a small, multi-purpose, maneuvering rocket that would work in conjunction with the station. It would retrieve satellites for repair and servicing at the station and would redeploy them when the work was completed.

A contract was awarded by NASA to TRW, Inc., during the mid-1980's to undertake development of an OMV. The craft was to be equipped with a docking device and robotics to facilitate handling of spacecraft and provide for stationing the OMV at its own servicing port at the Space Station.

After several years effort and expenditure of more than $200 million, the effort was abandoned. Requirements for an OMV were trimmed from space station plans. At some time in the future, when the economics of servicing spacecraft become favorable, the OMV may reappear again as an operational requirement.

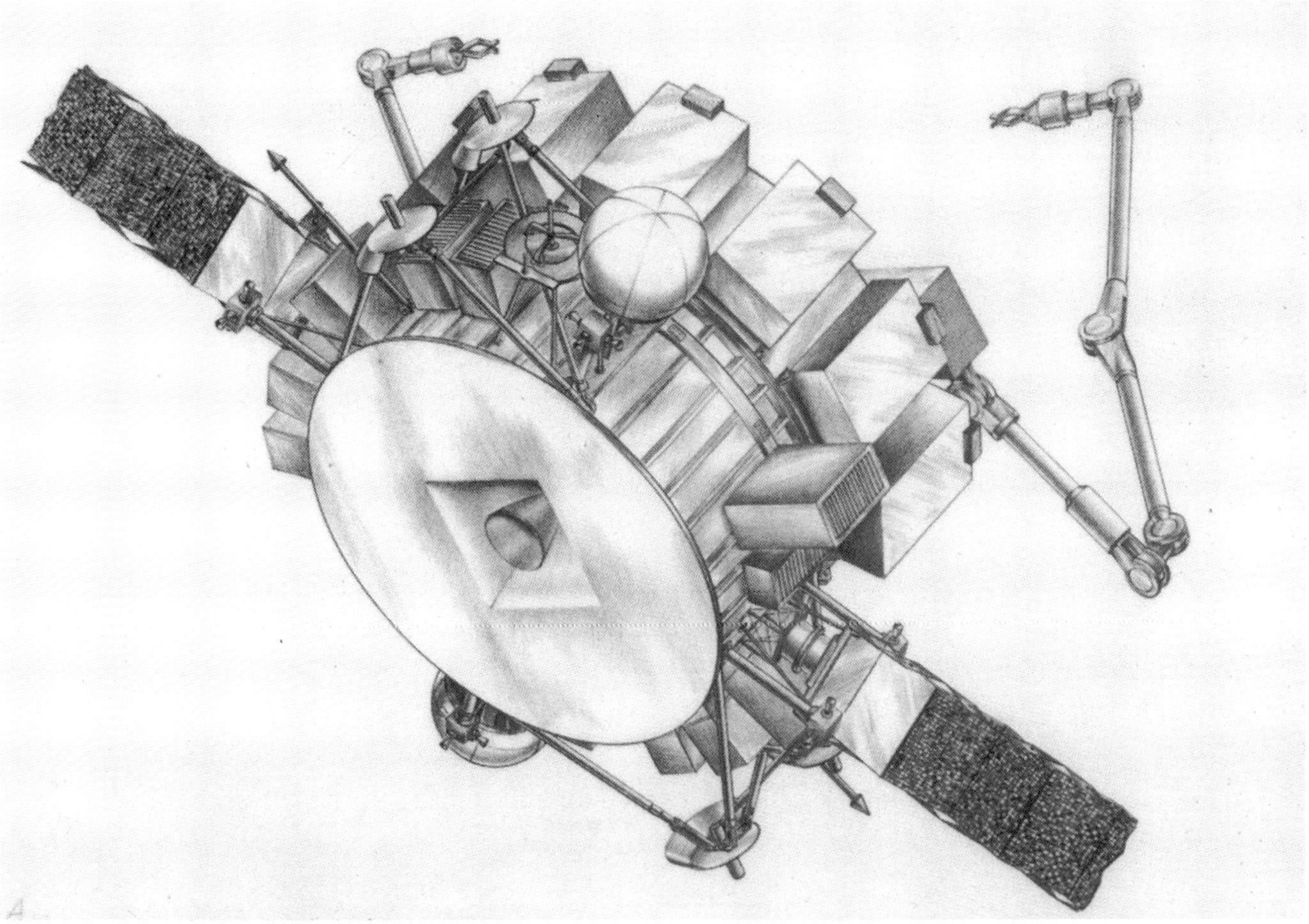

Sketch of robotically equipped orbit maneuver vehicle (General Dynamics)

14-6 Outlook

Appearance of a budget line item for an advanced version of Centaur would be a decided step toward evolution of this vehicle into eventual service as a reusable Orbit Transfer Vehicle. Such a commitment by the government appears to have been made in 1993, but authorizations by NASA to start work have not been made. Were this work to be accomplished, Centaur will have taken a big step forward in improved methods and processes, lower cost of production and better reliability. Centaur is the only high energy stage in the U.S. inventory. A program of continuous product improvement makes good sense, both from the standpoint of avoiding product obsoletion but also to be in readiness for more ambitious space missions, should they be required in coming years. The advanced Centaur competes effectively with the best that designers have been able to create in all-new OTV concepts.

Solid rocket motor technology appears to be well-geared to the requirements of contemporary spacecraft. In many cases it provides the cheapest and simplest way to place satellites into their desired orbits. Competition between the companies that serve the satellite manufacturers keeps prices in line and encourages innovation. The picture is less secure for systems such as Orbital Science's TOS and Boeing's IUS, where cost of production has all but removed the competitive edge against liquid fueled upper stages. These systems started out with the promise of low cost operations but fell short of the goals that were initially set.

The OMV remains a vehicle of the future, in an era when manned activity in near-earth orbit encompasses routine servicing of satellites, as well as orbital basing of transfer vehicles. The infrastructure for such operations has been thought about and studied, but no plans exist to undertake such operations, perhaps for the next two decades.

The hybrid orbit transfer vehicle remains an engineering curiosity, looking for a home. Should the Air Force proceed with plans to develop a new family of expendable launch vehicles, replacing Atlas, Delta and Titan, a hybrid upper stage would be useful for missions at the lower end of the payload range over which the system is planned to operate. It offers many features that are desired for rapid availability, simple operations, low cost and good reliability.

15

Advanced Rocket Concepts

15-1 Futuristic Thinking

For many years scientists and engineers all over the world have dreamed of and labored at extensions of technology that would enable huge jumps in human capability to explore and exploit the challenges and mysteries of space. These technology extrapolations center largely on advanced propulsion and super-robotics, the latter of a type that is self-replicating and self- repairing. The subjects of science fiction often stem from more learned conjectures regarding stellar and planetary objects and the role of humans in the cosmic picture, should they dare to venture away from the home planet. Somewhere along the way science fiction and reality merge, and where they merge will be a direct consequence of technology advancements and human daring, and perhaps human need.

The dreams take different forms, some confined to our own solar system and others bent on interstellar travel to learn more about what goes on beyond our system.

During the 1970's, the British Interplanetary Society undertook a study under the name "Project Daedalus," which would send an unmanned starship to Barnard's Star, a small red dwarf, situated in the constellation Ophiuchus, six light years distant. The project, headed by Alan Bond, examined technology extrapolations that would permit undertaking a venture of this nature. The technical requirements were bounded by a desire to accomplish the mission within the lifetimes of the youngest participants in the project, which in turn established a requirement for a major advancement in propulsion. The time frame for undertaking the trip was estimated to begin about a hundred years from now, largely on the premise that Helium 3, needed for fuel, would be in plentiful supply. This in turn was premised on fusion power supplying most of the Earth's energy needs, with Helium 3 being "imported" from Jupiter, where it is present in quantity.

The future science of global terraforming was kicked off in September of 1948 when the eminent mathematician John von Neumann gave a lecture entitled: "The General and Logical Theory of Automata." The substance of von Neumann's interest was to lay the foundation in strict mathematical terms for the theory and functioning of immensely complex automated machines. A first conclusion of his work was that self-reproducing machines can in principle be built. In one of his stories, science fiction writer Isaac Asimov took up the problem of supplying Mars with water from Saturn's moon Enceladus, which has a mass equal to about five percent of the Earth's oceans, with automated rocket freighters. Physicist Freeman Dyson, of the Institute of Advanced Studies at Princeton, has in his "thought experiments" extended Asimov's ideas into self-replicating transporters that would move water from Enceladus to Mars on a huge scale. This may be the stuff of futurists, but it is no longer difficult to imagine fleets of robot transporters, self-replicating, operating like cosmic ants between the resource-rich moons of Jupiter and Saturn, and the planet Mars, stocking the planet with the elements and compounds needed to sustain a comfortable living environment. Global terraforming has caught the attention of many long range thinkers who are looking toward preparing other planets for human habitation. Some of these plans seem to leap beyond the bounds of human capability, but perhaps they are intended for other generations, driven by the requirements of their own time.

Gerard K. O'Neill, of Princeton University, is a proponent of large space colonies situated in high-earth orbit, each capable of sustaining tens of thousands of humans as comfortably as their earth-bound counterparts. Gerry and his team of investigators plan to tap the moon as the source of materials of construction for the space colonies. For their purposes, rockets would be too costly to be employed as a means of materials transport. Instead, the bulk of the material would be launched toward the orbital construction sites via electromagnetic accelerators located on the surface of the moon. Nevertheless, a venture of this magnitide would require a large space-based and lunar-based supporting rocket fleet.

Peter Glaser of Arthur D. Little, Inc., has promoted the concept of supplying a significant part of the Earth's electrical power requirements from space-based solar power collectors. Huge arrays of solar cells, located in geosynchronous orbit would generate electrical power. Converted to microwave energy, the system would beam to collectors on the Earth's surface which, after reconversion to ac power, would distribute it to users. Others have proposed the same kind of system, but situated on the surface of the moon, arguing that the logistics of construction of large arrays are simpler in that approach. Either way, an advanced rocket-based transportation infrastructure would be required to accomplish installations of this nature.

Such ambitious plans move rocket designers to advance their art, seeking methods of propulsion with efficiencies that far exceed that of conventional rocket systems.

The Orion launch concept, originated by scientists at General Atomics Division of General Dynamics in 1958, was terminated in 1965, two years after the international test ban on atmospheric nuclear explosions. Cancellation was fully justified as Orion was a "dirty" rocket, leaving nuclear refuse wherever it went. But investigative work along the same lines continued at Los Alamos and Lawrence Livermore laboratories. The idea of repeated nuclear pulses operating against a spring mounted "pusher plate" was retained, while more advanced methods of achieving nuclear detonations were sought. The impetus for continuing such work lies in anticipation of possible future requirements related to fusion power. Helium 3 is involved in one of three likely combinations of deuterium, tritium and helium for fusion power generation. While rare on Earth, it occurs in abundance on other bodies in the solar system. Consumption of He3, based on supplying the world's power needs via fusion power, could exceed 1000 tons per year. Pulsed nuclear powered rocketry is a favored way of achieving transport capability for a He3 supply system. Some analyses predict specific impulses four orders of magnitude higher than chemical propulsion.

The Daedalus project engineers settled on a pulsed fusion rocket system for the Barnard's Star probe, relying on studies that had been done at Los Alamos and Lawrence Livermore in obtaining small fusion explosions by laser or electron beam ignition, as well as work by Winterberg, proposing use of relativistic electron beams to initiate fusion explosions and a concave mirror to direct the particle fluxes.

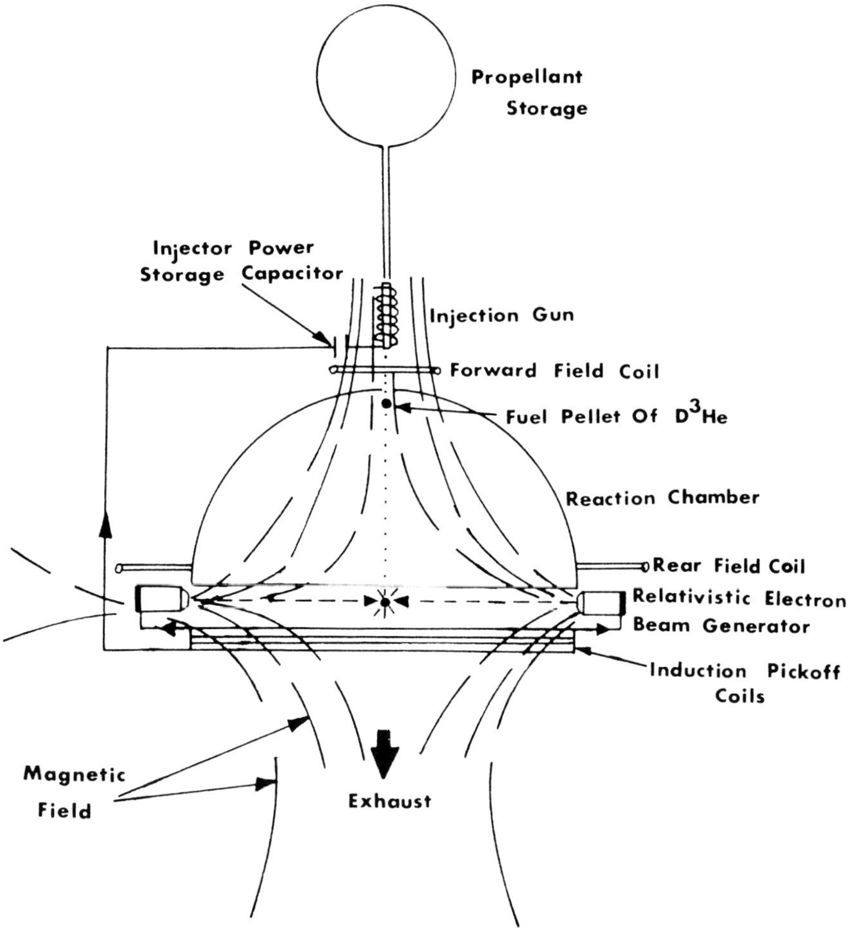

A schematic diagram of the Daedalus propulsion system (British Interplanetary Society)

15-2 Relativistic Velocity Rockets

The ultimate space vehicle, at least as far as present day physicists are prepared to admit, is the photon rocket. In a photon rocket, an electromagnetic beam is used to produce thrust, resulting in an exhaust velocity equal to the speed of light. Compare this with the exhaust velocity of a chemical rocket engine, at only five to six thousand feet per second. But a photon rocket lies considerably beyond present day technologies. For every pound of thrust, the rocket would require over a hundred MW of power. Were it possible to do this at thrust levels sufficient to generate an acceptable acceleration level, the next problem would be to invent an engine with extremely high reflection efficiencies, to prevent destruction of the vehicle. Ordinary mirrors would not work, so some other type of mirror would have to be devised.

The key to creation of a photon rocket probably lies in learning how to work with antimatter, or antihydrogen. The subject has been under investigation for some time by scientists around the world. Antimatter represents the highest possible energy storage density known to science. Antimatter converts all of its mass into energy when it comes into contact with normal matter, and therefore must be kept in a near perfect vacuum, suspended by electrostatic or electromagnetic fields. So far, it has been created only in minute quantities in high energy "atom smashers."

In principle, antimatter can be used directly in a rocket, using it to heat a propellant to achieve very high exhaust velocities. Rockets could then be made much smaller for conventional missions, or could be designed for new, high-velocity missions. An aerospace plane weighing 120 tons at liftoff could carry 20 tons of payload into orbit using only 35 milligrams of antihydrogen, plus several tons of hydrogen as the propulsion medium. But estimates of costs of production and storage of antihydrogen, using near-term methods, could cost $10 million or more per milligram. Hence it is clear that antimatter has nothing to offer in terms of reduced cost of rocket launches by present day standards.

Another advanced concept centers on an "interstellar ramjet." This type of rocket carries very little fuel, but instead scoops up hydrogen gas from interstellar space, feeding it to fusion reactors to produce power when the nuclei of hydrogen are fused. The ram scoop is extremely large, measuring thousands of kilometers across, with the spacecraft at its center. The gas entering the ramjet would first be ionized by means of lasers, then swept up to a collection point by a magnetic field. The technical problems related to interstellar ramjets are enormous and by no means is it clear that losses at the ramjet inlet will be acceptable. An interesting aspect of a space ramjet is that its development would necessarily be based in space, since in this case the designers and builders would be forced to work where the propulsion medium is located.

15-3 Unconventional Propulsion

Several unconventional launch methods have been proposed that are non-rocket in nature. Whether they offer advantages over rocket propulsion is as yet unclear, but low-level funding from different sources tends to keep the investigations going.

One method called laser propulsion, generates propulsive force by beaming a high-powered laser from the ground to the launch vehicle. The laser beam follows the vehicle during its ascent, heating a working fluid such as hydrogen or water, which is the propulsive medium. Laser pulses would create a detonation wave of hot, expanding vapor, which would produce thrust. In laser propulsion, the working propellant is not necessarily a fuel, since there is no combustion. One concept considers water ice to be an ideal propellant. Many technologies need to be worked to expand this idea, including ground power sources, thruster efficiency, beam quality, atmospheric propagation, guidance, and environmental effects. Like many advanced concepts, it is difficult to conceive how such a system could undergo development testing.

Another propulsion method, useful only for transportation of cargo because of the high accelerations involved, is the Ram Cannon. In this device, a barrel is filled with gaseous propellant and the projectile, which flies through the propellant, igniting it like the centerbody of a ramjet engine. An experimental ram cannon has accelerated a one pound projectile at 20,000 g's to a velocity of 1.25 miles per second, which is about 20 per cent of the velocity required to reach earth orbit. A full scale cannon would have a barrel approximately two miles long. Such devices are believed to be useful for launching propellants into orbit, which constitute more than half the current U.S. payload mass-to-orbit. Little thought has been given to what transpires at the "catching" end of the trajectory, where one encounters the problem of how to manage hundreds of small vessels containing propellants, including disposal of the empty containers.

Electromagnetic devices for launching payloads into orbit have also been studied recently. These, too, are high acceleration devices limited to inert cargo transport. A vertical electromagnetic coil-gun in an 8 kilometer deep hole is said to be capable of launching a one-ton payload into orbit.The cargo-projectile is launched by successively activating electromagnetic coils along its acceleration path. The coils for such a launch would require ability to store and deliver the output of a 1000 megawatt power plant in 90 seconds. The major problems in coil guns, switching, and power storage, are claimed to have been solved by at least one company working in this area.

15-4 Alternative Nuclear Powered Rockets

Since the performance of solid core, nuclear rocket motors is limited by the temperature at which such devices can operate, it is to be expected that scientists and engineers would seek ways of operating at higher temperatures through innovations in reactor design. One such design proposed in 1960 is a rotating fluidized bed reactor, which offers lower temperature gradients in the nuclear fuel than in solid core designs.

Modest development work on the rotating bed reactor was carried out for over ten years. In this design, the carbide nuclear fuel, in the form of 100-500 micrometer diameter spherical particles, is maintained in an annular configuration by centrifugal force in a cylindrical container. The cylinder is of a porous material and is rotated by an electric motor. Hydrogen propellant passes first through the nozzle to cool it and then through a beryllium reflector. The hydrogen, now gaseous, next flows through the cylinder wall and into the annular core at a velocity sufficient to fluidize the nuclear fuel particle bed. The heated hydrogen next flows out through the nozzle, producing thrust. Control of the rotating bed reactor is achieved by means of drums in the reflector. The control drums contain a moderator on one side and a neutron absorber on the opposite side.

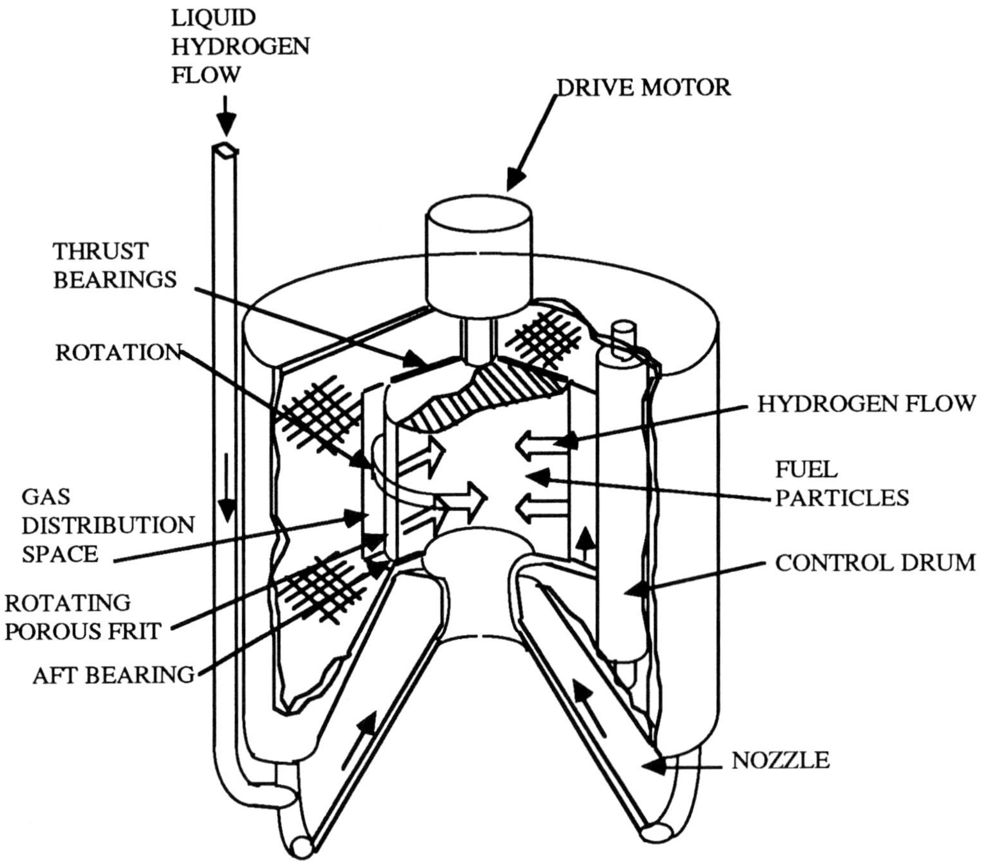

Diagram of a rotating fluidized bed nuclear rocket engine

Gaseous core reactors have also come on the scene as a propulsion device with considerably better performance potential than solid core nuclear rockets. A gas core reactor contains the nuclear fuel in a gaseous phase in an externally moderated enclosing cavity. The gaseous material might be uranium hexaflouride or, if temperatures can be elevated, vaporized uranium metal can serve as the nuclear fuel, existing as a fissionable plasma. Operation at elevated temperatures is possible by keeping the plasma separated from the cavity walls hydrodynamically through vortex confinement techniques. The crux of the technology is concerned with separation of the plasma from the cavity walls while keeping the losses of fuel to a minimum. Through this technique, the operating temperatures can be considerably higher than the cavity materials of construction can sustain.

Technology advancements such as these yield twin benefits of lighter weight for the engine, as well as a specific impulse that may be double that of the Rover class of reactors. Many other innovative nuclear reactor concepts have been proposed for advanced space propulsion, but experimental work to support the ideas is limited or nonexistent.

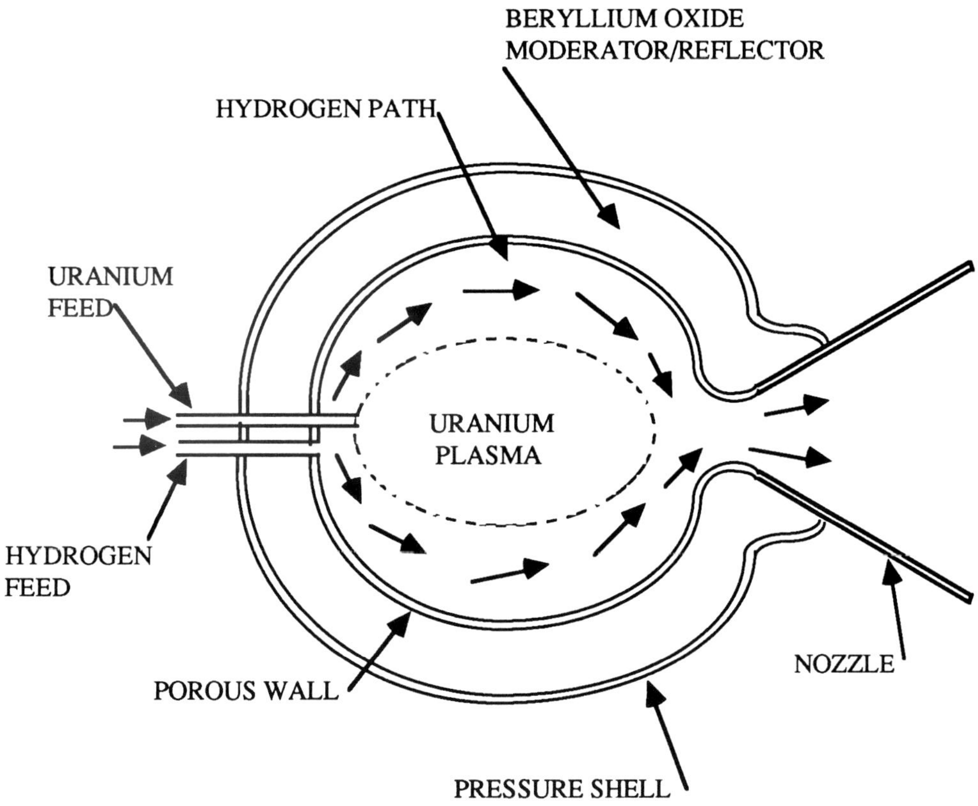

Diagram of a plasma core rocket motor

15-4 Electric Propulsion

Electric propulsion has been under development, mainly by NASA, for over three decades, as a potential economical means of transporting cargo across space. The very high specific impulses available in such devices translate into a much reduced requirement for rocket propellant, which, of course, has to be transported into orbit via chemically powered rockets. A main drawback to electric propulsion is the very low thrust levels that can be generated, which results in long transfer times. This limits their usage to missions where time is of small consequence in the overall plan. Electric propulsion would be unacceptable for transporting communications spacecraft to geosynchronous orbit due to lost time that could be generating revenue.

To reach geosynchronous orbit or escape velocity, an electric OTV will gradually spiral out from a low-earth orbit, constantly thrusting. Only a small elevation is achieved with each orbit, so the process can take many months to complete. Chemically propelled transfer vehicles, on the other hand,, acquire transfer velocity in a single burn, lasting only a few minutes, then coast the remainder of the trip over a six hour elapsed time. As a consequence of the long transfer periods, electric OTV's suffer a greater gravity loss than chemical rockets, and therefore do not deliver the total performance that one would expect as a result of differences in specific impulse.

Four types of electric propulsion are candidates for operation of an OTV: resistojets, arc jets, magnetoplasmadynamic (MPD) accelerators, and ion accelerators. Resistojets have found an application as control thrusters for spacecraft and have not been seriously considered for orbit transfer purposes. Arc jets and MPD thrusters are generically similar, so the two types of electric propulsion devices that will be described are ion and MPD accelerators .

Ion thruster development has been centered mainly at NASA Lewis laboratories and has been operated successfully on experimental spacecraft. Early versions used mercury for propellant, but concern about mercury contamination of the atmosphere, as well as unavailability of mercury for large scale space operations, led eventually to argon as the preferred propellant. Argon is both plentiful and cheap and can be handled like liquid oxygen as a deep cryogen.

The most serious MPD investigations have been undertaken by Jet Propulsion Laboratories and Princeton University. MPD thrusters operate at a much higher thrust level than ion thrusters, but have an overall efficiency of about 50 % as compared to 80% for ion thrusters. Low efficiency translates into heat loss, which means that engines of this type may have to be equipped with large radiators. Alternatively, OTV's could

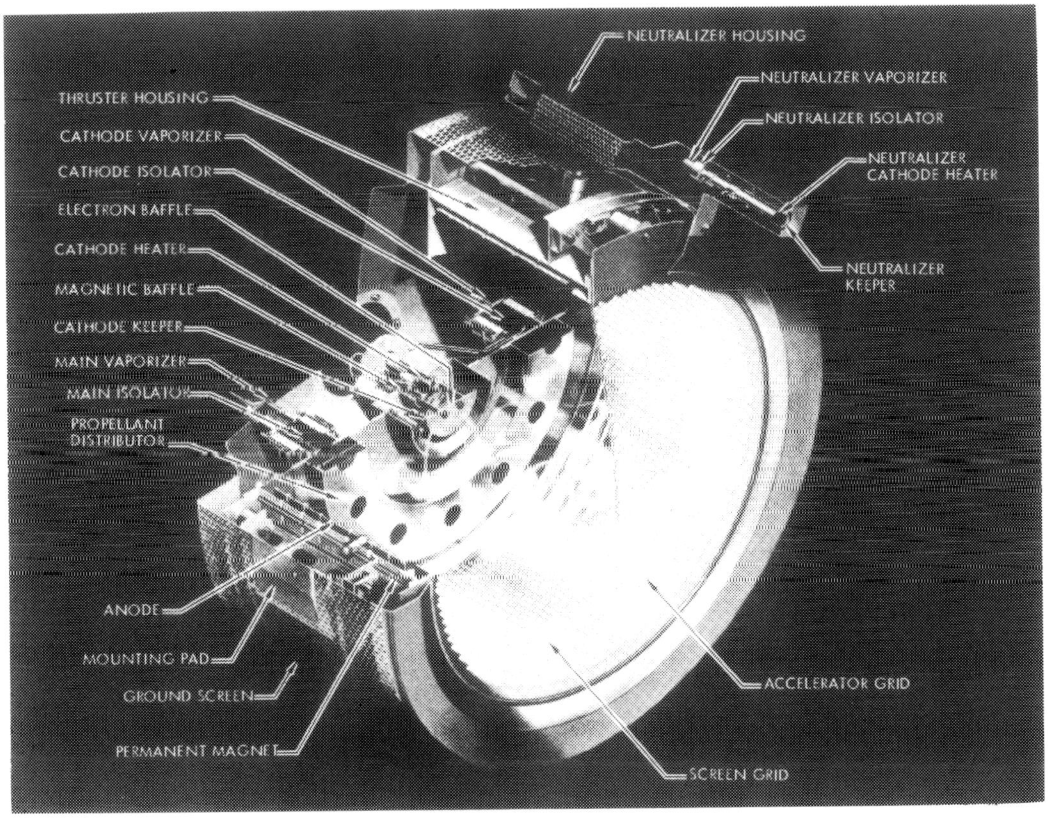

Cutaway drawing of an electron-bombardment ion thruster (NASA LeRC)

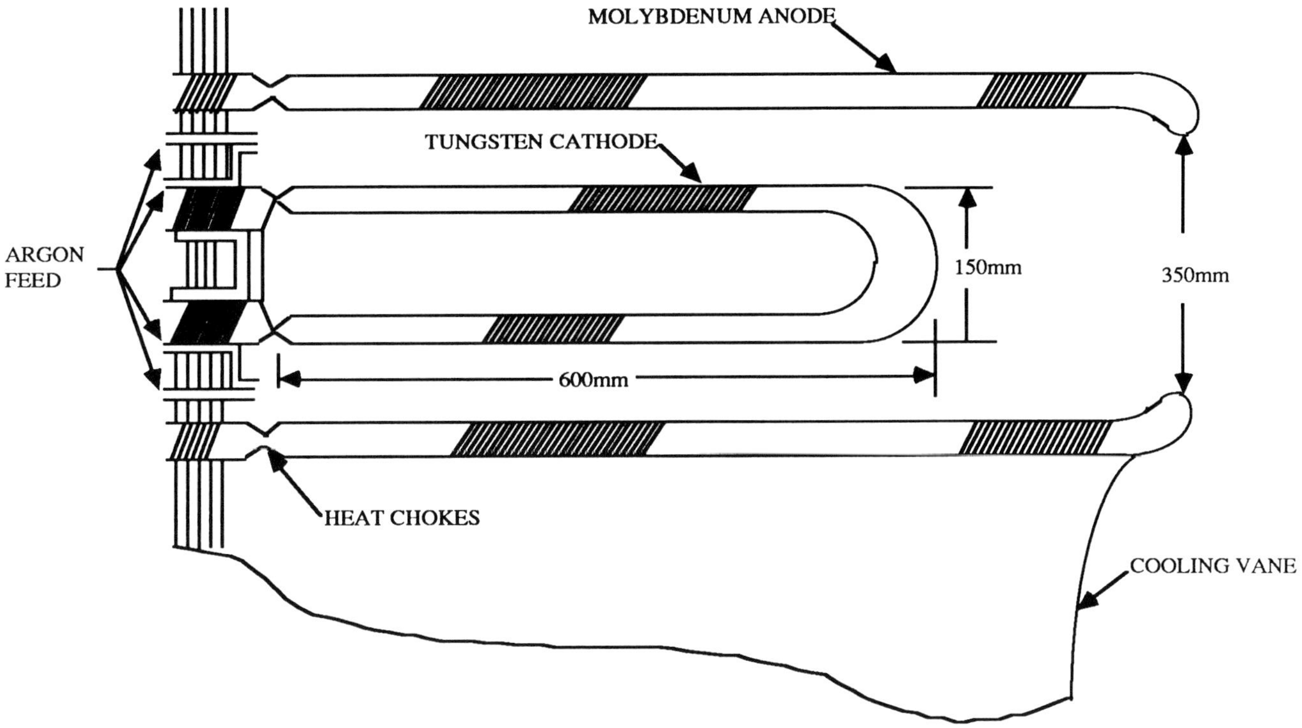

Diagram of a magnetoplasmadynamic thruster

be equipped with many engines which would operate in a pulse mode, taking the time between firings to cool by self-radiation.

MPD thrusters operate at very high current densities. A typical device, capable of producing ten pounds of thrust, would operate at about 200 volts and a current from 37,500 to 50,000 amperes, and thus would require a 10 MW power supply. Power at this level can be provided either by nuclear reactors or by large solar arrays. The problem of heat radiation for a continuously operating engine would be formidable if fifty percent efficiency is the best that can be attained.

A broad variety of nuclear reactor designs has been created in the past to meet potential space power requirements, as well as propulsion requirements. The programs have included Rankine cycle liquid-metal cooled fast reactors, thermionic reactors, fluidized bed reactors, and gaseous core reactors. Solar arrays, however, have provided the bulk of the power requirements for spacecraft to date, although sizes do not approach anything like the requirements of an electric OTV. During the 1970's, however, design studies of the Solar Power System (SPS), funded by NASA and the Department of Energy, accomplished conceptual designs of solar arrays measuring tens of square kilometers in area.

The accompanying illustration of a pulsed MPD powered OTV based on solar power depicts a 5MW OTV that would be used to shuttle payloads to and from geosynchronous orbit. This huge space butterfly uses argon for propellant and is powered by fifty 5 pound thrusters that operate sequentially to permit radiation of heat generated during each pulse.

Despite the years of advanced development to date, both ion and MPD powered OTV's face a very large development program before a flight system can be put into operation. For a device such as illustrated, technology development has to be pursued along several lines, including lightweight solar arrays, lightweight, efficient support structures, dynamic control of large and fragile structures, high-power switching, lightweight power bussing, power conversion at megawatt levels, thermal heat rejection, and long-life MPD reactors.

Prospects for development of an electric OTV would seem to be blocked on the sole reason that such devices can accomplish only a limited part of a mission plan; those segments that are not time dependent. A chemically powered or nuclear powered OTV can, on the other hand, take on all missions. The chief consideration is whether the cost of transportation of propellants to low-earth orbit for fueling OTV's is an overriding factor in the choice of systems. So far, economic analyses have failed to turn up such a conclusion.

FAST ELECTRIC TRANSFER SYSTEM (FETS)

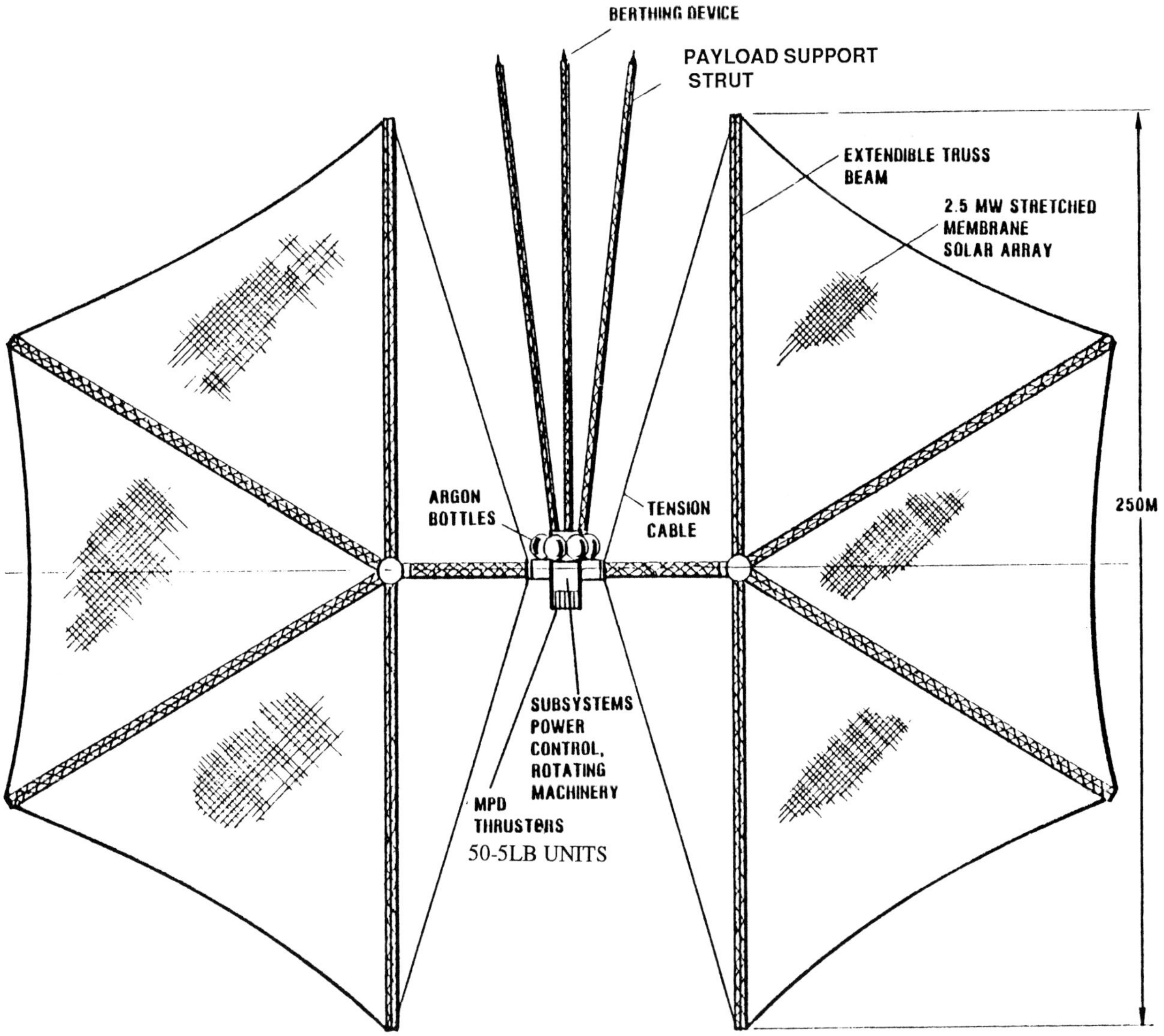

A solar powered orbit transfer vehicle, using MPD thrusters with argon propellant.

15-5 Outlook

Advanced propulsion systems await the next major space initiative, whether it be exploration of Mars, establishment of bases on the moon, development of solar power systems for beaming electrical power to Earth, extensive exploration of the outer planets and their moons, or beginning the mining of asteroids. All of the above events will probably occur, but the current state of U.S and European economies provides no basis for optimism regarding their occurrence, or even initiation in the next twenty five years. Nevertheless, research will continue in all areas under DoD, NASA, and DoE sponsorship, bringing a refinement to the methods already suggested, as well as new innovations that will be meaningful to future space systems.

A powerful medium for accomplishment of such studies is through the government's Small Business Innovative Research (SBIR) program, where small companies can be granted research contracts up to $100,000 each for exploring new ideas, with additional grants up to $500,000 for those that show merit. All that is needed to stimulate studies in advanced propulsion is for the SBIR sponsors to insert appropriate requirements in the form of investigative titles into their annual solicitations for proposals.

As mentioned in chapter 12, steam rockets may be the final solution for a general purpose rocket powered space transfer system. Should future studies confirm this to be so, development of high temperature reactors will doubtless take precedence over other types. The plasma core reactor described on page 185 points the way to successful development of an efficient steam rocket. High temperature reactors would allow steam rockets to approach the same performance that is achievable with hydrogen, using Nerva type reactors. The steam rocket would have the advantage in cost and simplicity.

16

Is There a U.S. Space Policy?

16-1 Space Advisory Board

Over the past four years, several high-level committees have been convened in Washington to assess the future of United States space ventures and the launch capabilities required to support them. Notable among these was the Augustine Committee, charged with developing an overview of U.S. space efforts well into the twenty-first century. The most recent was a 1992 task group effort commissioned by former Vice-President Quayle, chairman of the White House Space Council, directed toward review of the nation's overall space-launch strategy, as approved by the President in July of 1991.

This strategy, delineated in National Space Policy Directive 4, had four major elements: to ensure that existing space launch capabilities, including support facilities, are sufficient to meet U.S. manned and unmanned space-launch needs; to develop a new unmanned, but man-ratable space-launch system to reduce costs and improve performance; to sustain a vigorous space-launch technology program that would apply to both existing and new space-launch systems; and to actively consider commercial space launch needs and factor these needs into the decisions on improvements in space launch facilities and launch vehicles. The president directed that all affected government agencies implement these elements of space launch strategy within constraints of overall resource availability and other policy guidelines. The strategy was based on the conclusion that if the United States is to compete effectively in the future, it must take near-term actions that will improve the efficiency of its space-launch operations, maintain its reputation for reliability, and significantly reduce the cost of space launches.

The task group, chaired by former secretary of the Air Force E.C. Aldridge, was faced with the issue of determining whether the conditions leading to the strategy continue to be relevant in today's environment. Its charter group produced a set of recommendations in a report filed in November of 1992 and urged immediate implementation.

The panel's findings included an observation that "the National Space Launch Strategy is not being implemented in a cohesive, coordinated and integrated manner by the affected government agencies. Inconsistency in planning, lack of a formal coordination processes for integrating plans and programs, and lack of program definition and priorities are indications that the management of the implementation of the strategy across government agencies has been inadequate."

The committee confirmed that near-term launch needs, through the year 2000, are being met by existing systems. Improvements are under way in virtually all launch systems and launch infrastructure. Even so, the task group concluded, the system is fragile, not as reliable as it should be, expensive to operate, and lacking operational flexibility.

These factors reduce U.S. ability to be competitive with foreign launch vehicles and inhibit reaping the benefits of space due to the high cost of launching payloads. A key factor, in the committee's view, is that the current launch vehicles, orbit-transfer vehicles, and launch-support systems are based on the older technologies that date back to the 1950's when the U.S. entered the space exploration era. Heavy penalties are incurred by frequent launch delays that stem from repeated tests to guarantee success on vehicles known to be no better than 95% reliable.

The Task Group addressed the subject of current U.S. overcapacity in space-launch vehicle production, based on projections of government and commercial requirements. Besides the Space Shuttle, the United States has three major launch-vehicle systems in production; Atlas, Titan and Delta, as well as six or seven in development or production for the small payload market. This overcapacity, as well as entry of launch vehicles from the nonmarket economies of Russia and China into the competitive arena, signals a need for downsizing the launch-vehicle industry as well as a need to develop a launch vehicle that will be more reliable and resilient than existing systems and competitive in the world market.

The committee deplored the high cost of shuttle launches, but recommended only that solutions for the high cost of operations must be found. The solution proposed in Chapter 5 would abandon the notion of the Shuttle as a space transportation system, treating it instead as a returnable, reconfigurable space station. Putting the Shuttle into a space station budget is an appropriate way to solve the Shuttle's dilemma as a very high-cost launch vehicle.

While the panel's analyses and evaluations were in progress, the Congress approved the Defense Appropriation Bill that terminated National Launch System development effort for Fiscal year 1993, appropriating only $10M in the NASA budget to continue NLS technology. This confirmed, with unexpected emphasis, previous concerns regarding lack of integration and coordination in U.S. space policy.

The committee's recommendations reflected its findings regarding mission needs, industry overcapacity, requirement for a new economical launch system, transition plan to the new launch system, commercial launches, and overall space policy, followed by a strong response to the termination of the National Launch System effort. The panel did not take issue with the termination of the NLS since the thrust of this program was to develop a family of launch vehicles that would ultimately lead to a heavy-lift launch vehicle. A near-term requirement for such a vehicle was rejected by the task group, since nearly all government and commercial payloads in the near future can be launched with a vehicle in the lower range of the capacity planned for the NLS program. In its place, the panel recommended development of a new launch vehicle with a payload range between 20,000 pounds and 50,000 pounds to low-earth orbit, together with a new cryogenic upper stage for high-energy missions.

The committee recommended changes that it felt would result in better planning and coordination of U.S. space-launch capability. An executive committee consisting of heads of the major agencies involved in space launch should be formed to provide overall space-launch program and planning guidance. A further recommendation was to designate a single space launch authority, who would be responsible to the executive committee for planning, coordinating, and integrating U.S. space launch capabilities. This individual would be an executive-level appointee.

The proposed new launch vehicle is called the Spacelifter. As then conceived, it would employ applicable NLS technology and operational concepts that hold promise of reducing costs and increasing system reliability. It would be compatible with both manned and unmanned payloads, and have a performance capability that ranges between 20,000 pounds and 50,000 pounds to low-earth orbit. Unmanned operation would come first. The system would eventually be man-rated to enable launching of a future personnel carrier.

With respect to the Space Station, the panel recommended a two-phased space-launch program. The first would continue to use the Space Shuttle for deployment and man-tended phases of the Space Station. A specific statement was made against creation of a heavy-lift vehicle for deployment of the station. The second phase would use a Personnel Launch System (PLS) and a cargo transfer vehicle for support of ongoing operations, replacing the Space Shuttle. The launcher would become the primary, long-term support vehicle for the Space Station.

Establishment of a small payload launch program was recommended to encourage space research and experimentation aligned to capabilities of such vehicles, which offer significant cost savings. The committee believed the government should establish a central office that would better plan, integrate and coordinate government-wide efforts for this class of vehicle.

Although the committee's recommendations were thoughtfully presented, they had little impact on the Congress. TheWhite House Space Council to whom the report was directed was disbanded by the present administration. Attention has since been diverted to studies headed by General Moorman in 1994, regarding new launch vehicle development.

In 1963, at the dedication ceremonies of the new General Dynamics Astronautics facility in San Diego, California, a time capsule was buried containing predictions relating to space and the state of humanity a hundred years hence. The time capsule is to be opened in the year 2063 A.D. for the amazement of the citizenry of that period. The predictions were solicited from persons then held in high esteem for their achievements in U.S. Government, academia, and the Department of Defense. Thirty years have passed since the capsule was sealed, almost a third of the time to the year 2063. It is interesting to see what progress has been made against the visions held by those notables.

Dr. William H. Pickering, President of the American Institute of Aeronautics and Astronautics, predicted that mankind will be travelling at relativistic velocities to nearby stars with the purpose of exploring other planets. He also foresaw permanent scientific colonies at various stations throughout the solar system. Based on progress to date, Dr. Pickering's look into the future was perhaps too optimistic. However, in the event that the SETI program reads signals someday from a not-too-distant star, the technology that would permit sending an unmanned probe to that region could be well underway by 2063.

Dr. Edward Teller foresaw use of ballistic missiles to transport materials over big distances in a short time, but discounted personnel travel by this method as being too uncomfortable for the typical traveller. Teller's prophesy could be fulfilled in the next ten years. Development of a reusable launch vehicle similar to McDonnell Douglas' Delta Clipper rocket, but equipped with scramjet propulsion, will make rapid point-to-point transport over the Earth's surface both feasible and economical.

Fred Whipple, Director of the Astrophysical Observatory, envisioned the planet Mars responding to global terraforming, its atmospheric pressure and oxygen content having been raised so outdoor work can take place with simple face masks. Mars is able to raise all food for its 700,000 inhabitants. Another prediction, based on SETI having picked up intelligent radio sources from the Tau system, where two dwarf stars coexist only a half light year apart, sends an unmanned probe toward Tau in the year 2045. One vision almost matched that of Dr. Teller, except that personnel travel is accommodated by the technique of constant acceleration and deceleration over the flight path.

The Honorable George C. Miller, Chairman, Committee on Science and Astronautics, Congress of the United States, predicted that humans will have viewed Mars from close at hand. George Miller's views must have been tempered by a close acquaintance with the workings of Congress, which in the final analysis sets the pace of development by the level of funding it approves. Judging by the current rate of progress with regard to Mars exploration, mid-century 2000 may be a fair estimate for the timing of the first manned visitations to Mars.

Among eleven separate predictions, Krafft A. Ehricke, Director of Advanced Studies, General Dynamics/Astronautics, identified routine earth-to-orbit and global transportation with space-plane type aero-chemical vehicles. The National Aero-Space Plane project, begun just a few years ago, indicates that this forecast is on track. Ehricke also foresaw application of large reusable rockets for point-to-point cargo transport on Earth. Again, the Air Force sponsored Delta Clipper, now in early development at McDonnell Douglas, may be the first step in that direction. Dr. Ehricke also predicted that controlled thermonuclear reactors would be the primary propulsion devices for interplanetary vehicles, reducing the flight time to Mars or Venus to less than 90 days and trips to the outer planets to less than six months. Present thinking regarding future planetary missions appears to favor nuclear propulsion.

Based on current programs and achievements of the past thirty years, it can be concluded from these samplings that the more conservative predictions will have been vindicated by the year 2063. But progress in space is not hindered by scientists, engineers, and U.S. industry. This combination is capable of undertaking any technical challenge. It is the absence of dreamers among the nation's planners and policy makers in Washington that can slow the space program to a crawl.

16-3 Concluding Thoughts

Space exploration and space research are important for several reasons, not the least of which is that they open the door to space *exploitation*, which leads to direct benefits to humankind. In this context, they have revolutionized communications. Geosynchronous satellites now connect the world for telephony, data exchange, and entertainment. Earth observation satellites enable accurate tracking and forecasting weather, analysis of Earth's resources, assistance to farmers regarding crop management, and future planning of terrestrial changes. Specialized satellites like the Global Positioning System enable precise location of any object carrying a receiver such as ships at sea or aircraft. Military satellites provide Earth observation and information transfer essential to national security. In slightly more than four decades, earth-oriented satellites and their associated ground systems have become a leading industry, with enormous growth possibilities. Scientific advancements keyed to miniaturization of hardware have led to a myriad of consumer products that find a worldwide market.

Exploration begets beneficial exploitation, as it often has in the course of human history.

Moreover, space exploration adds to knowledge. It is one of several important lines of endeavor that is gradually gathering information leading to an understanding of who we are, where we come from, and why we are here. Purposeful, ongoing investigations along these lines must always be maintained and encouraged. The alternative is to retreat into a dark age that will leave our distant progeny wondering about what turn of events led to humankind's temporary loss of its adventurous spirit.

The current climate with respect to space exploration tends to meager efforts, placing at blame an economy that needs to give priority to other civilian needs. Of course, those problems will never be solved by retreating from space exploration, but it is difficult for most politicians to weigh the long-term benefits of space exploration as compared to the immediate demands of society. Repeated reports by independent advisory groups have outlined the paths the nation should follow in its future space ventures. Former President Bush himself undertook to begin a major effort toward manned presence on the moon and Mars through the Space Exploration Initiative (SEI). But the Congressional response can best be characterized as a numb disinterest that defied intrusion.

Insofar as the nation's future is concerned, there is a historical precedent for the consequences of retreat from the call of exploration. Centuries ago, China embarked on an ambitious exploratory effort over a large part of the Earth, long before the British, Portuguese, and Spaniards took to the seas. The nation had fleets of ships and skilled crews that covered a large section of the Pacific and Indian oceans, with visits to Africa and the Middle East. In the middle of the fifteenth century the Ming Emperors suddenly halted all voyages to foreign lands. The fleet was called back and the ships were burned. For the ensuing hundreds of years, this advanced nation existed in a state of isolation from the world, left behind while the west moved forward in science and technology.

There is no lack of dreamers among the scientists, engineers and other individuals regarding the tug of adventure in space exploration and research. But there are few dreamers among the policy makers and planners in Washington and none who can articulate the importance of space to humanity, now that it has the tools to carry on energetically with the business of space exploration. The industry itself is beset by dullards in its top management, whose vision for the future is sadly wanting.

U.S. rocketry can easily measure up to any space-venture requirements that may be placed upon it. What we will do in the future remains a function of the extent of conviction by the nation's President, by the members of Congress, and by aerospace industry leaders, regarding how important progress in space is to the future of our people. Ignorance breeds lethargy. A positive turn of events will require that they first be educated, which is just one more challenge for the nation's dreamers.

Bibliography

America on the Threshold - A report of the Synthesis Group on America's Space Exploration Initiative. Chairman, Lt. General Thomas Stafford. May 3, 1991.

International Reference Guide to Space Launch Systems - Stephen J. Isakowitz 1991 Edition, Published and distributed by American Institute of Aeronautics and Astronautics.

Launch Options of the Future - Special Report, prepared by the Office of Technology Assessment, Congress of the United States. July, 1988.

National Space Transportation System Reference - Volume No. 1 - Prepared by the National Aeronautcs and Space Administration, June 1988.

Report of theAdvisory Committee on the Future of the U.S. Space Program - Chairman, Norman Augustine. December, 1990

The Future of the U.S. Space Launch Capability - Task Group Report to the Vice President of the United States - E.C. Aldridge, Chairman - November, 1992

Payload User's Manual - General Dynamics Corp.

Project Daedalus - Journal of the British Interplanetary Society 1978

Prophesies of Distinguished Americans of Man's Employment of Space in 2063 A.D. - Limited printing by General Dynamics Corp., 1963.

Shuttle Derived Vehicles Study - Prepared by the Boeing Company, Space Division, Huntsville, Alabama, under NASA contract NAS8-32396, 28 July, 1977.

Space Nuclear Power - Joseph A. Angelo, Jr., Ph.D. and David Burden. Published by Orbit Book Company Inc. 1985

Space Shuttle Evolution Into the Next Century - Rockwell International (internal document) May 7, 1993

The Dream is Alive - Ernest W. Maurer, Ph.D. Published by Geosync Publications, 1991.

Disturbing the Universe - Freeman Dyson Published by Harper & Row,1977.

Threshold, Number Nine, Fall, 1992 - Published by Rocketdyne Division of Rockwell International

Journals and Periodicals, 1985-1994 *Space News, Space, Aerospace America, Ad Astra, Aviation Week, Spaceflight*

Index

P

Q

R

S

T

U

V

W

XYZ